MW00768578

Mahlke/Gössing Fiber Optic Cables

Fiber Optic Cables

Fundamentals
Cable Design
System Planning

By Günther Mahlke and Peter Gössing

4th revised and enlarged edition, 2001

Publicis MCD Corporate Publishing

Die Deutsche Bibliothek – CIP-Cataloguing-in-Publication-Data
A catalogue record for this publication is available from Die Deutsche Bibliothek

ISBN 3-89578-162-2

4th edition, 2001

Editor: Corning Cable Systems, Munich
Publisher: Publicis MCD Corporate Publishing, Erlangen and Munich

Printed in Germany

Preface

Since 1980 a new technology has become increasingly accepted in the field of communication transmission via cables. In contrast to copper cable technology, here signals are transmitted optically with the aid of optical waveguides – also called optical fibers. This development was supported by the availability of suitable semiconductor components such as lasers, light emitting diodes and photodiodes. At the same time, digital transmission systems already in operation were upgraded to meet the demands of optical fiber technology.

This book is intended to make the topic of optical cable technology and plant design understandable to a wider circle of readers. In order to achieve this, physical and chemical contexts are discussed, but without attempting to provide scientific precision of detail.

The fourth edition contains major revisions to include details of important new developments in the field of fiber optics. The part on the fundamental principles has been updated with an enlarged section on optical nonlinear effects and nonzero dispersion shifted fibers. The chapters on optical fiber construction and fiber optic modules and components have been considerably enlarged or rewritten.

This publication is directed towards all who deal with the design, construction and maintenance of optical cable plants. Furthermore, it will provide basic information in a field dominated by highly technical literature. In order to make the study of specialized publications easier, this book contains a detailed glossary of special terms.

As was mentioned in the previous edition, my coauthor, colleague and friend Günther Mahlke died in early 1995. It was tried to update the book while remaining true to the spirit in which it was begun.

Grateful acknowledgement is made to Dr. Bernhard Deutsch, who contributed his expertise to this edition.

Munich, March 2001

Dr. Peter Gössing
Corning Cable Systems

Contents

1 Historical Development

Light was used as a medium for communication transmission in earliest times already, e.g. in the form of signal fires. Reminders of this type of communication can still be found today, e.g. in the flag signals of the navy, in light houses, in traffic lights and in control lights on machines. About 200 years ago methods were considered for the transmission of information over longer distances using light. For example in France about 1790 Claude Chappe built an optical telegraph system; it comprised a chain of towers with moveable signaling arms. Information could be transmitted 200 km along it within 15 minutes. This system did not become obsolete until the invention of the electrical telegraph.

The American Alexander Graham Bell developed the photophone in 1880, whereby voice signals could be transmitted by means of light. However, this idea did not find a practical application because weather and visibility had a much too negative influence on the quality of transmission.

The English physicist John Tyndall suggested a solution to this problem in 1870, not long before Bell's invention. He demonstrated that light could be guided in a stream of water. His experiment utilized the principle of total internal reflection, which is also applied in today's optical fibers. Following the experiments in the area of light modulation by Bell and those on light guidance by Tyndall, it took until 1934 for the American Norman R. French to receive a patent for his optical telephone system. In it he describes how voice signals can be transmitted via an optical cable network, whereby the cables were to be made of solid glass rods or a similar material and should have a low attenuation coefficient at the operating wavelength.

Technical realization of this concept was only achieved 25 years later. First a suitable light source was found for use as a transmitter. In 1958 the Nobel prize winners Arthur Schawlow and Charles H. Townes developed the laser, which was first successfully operated in 1960 by Theodor H. Maiman.

The feasibility of manufacturing lasers from semiconductor material was recognized in 1962. At the same time receiver elements were developed in the form of semiconductor photodiodes. Now the remaining problem had to be solved – that of finding a suitable transmission medium.

In the beginning attempts were considered to guide light in mirror-surfaced hollow tubes by means of complicated systems of lenses. In England in 1966 Charles H. Kao and George A. Hockham suggested the use of glass fibers for light guidance. However, these glass fibers would have to show attenuation values of not more than 20 dB/km in order to construct useful optical transmission systems over appreciable distances, but in 1965 optical waveguides still showed attenuation of about 1000 dB/km. In the field of medical technology, however, glass fibers were already introduced in the 1950s for direct image transmission over very short distances.

In 1970 Corning Inc. manufactured step index fibers and achieved attenuation values of less than 20 dB/km at the wavelength of 633 nm. Optical fibers with graded index profiles reached 4 dB/km in 1972. Today attenuation values of less than 0.2 dB/km at 1550 nm are achieved in single-mode fibers. At the same time transmitter and receiver elements improved appreciably, both with regard to power and sensitivity as well as in durability. Corresponding cable technology together with disconnectable and permanent splice joints for optical fibers have made it possible to introduce this new transmission medium without problems.

The first optical cables were put into operation for telephone services in military applications on ships in 1973 in the United States. Western Electric tested the first system with optical cables in 1976 on their property in Atlanta. One year later the first field tests by Bell systems were carried out in Chicago over 2.5 km and by General Telephone in Long Beach over 9 km. Siecor Corporation – a joint venture of Siemens AG and Corning Inc. – was the first manufacturer to deliver single-mode fiber cables to a telephone company in New York in September 1983. In 1988 the first standard fiber optic submarine cable (TAT-8) was successfully laid across the Atlantic.

Siemens started with the first fiber optic cable test route 2.1 km in length in Munich in 1976. The first optical fiber link of the Deutsche Telekom AG was installed by Siemens in Berlin in 1977. Since 1978 worldwide applications of the new technology began using initially multimode fibers. To date millions of kilometers of single-mode fibers in Corning cable technology are being installed annually all over the world.

2 Physics of Optical Waveguides

2.1 Electromagnetic Spectrum

Electromagnetic waves have proved to be useful for transmission of information for more than a hundred years. This is due to the fact that a metallic conductor is not absolutely necessary as a means for their propagation. On the contrary, they can also propagate at high speed in a vacuum or a dielectric medium, i.e. in an electric nonconductor.

An overview of the spectrum of electromagnetic waves and their applications is provided in Figure 2.1. Visible light occupies only the narrow range from 380 (violet) to 780 nm (red). This range is bordered at the lower wavelengths by ultraviolet radiation and at the higher wavelengths by infrared radiation.

In optical telecommunication with optical waveguides the near-infrared wavelength range of 800 to 1625 nm is used. For practical purposes the operating wavelengths for fiber optic transmission systems have been divided into five windows:

First window	around 850 nm
Second window	around 1310 nm
Third window	from 1530 nm to 1565 nm (Conventional or C-Band)
Fourth window	from 1565 nm to 1625 nm (Long wavelength or L-Band)
Fifth window	from 1380 nm to 1525 nm

Electromagnetic waves propagate in vacuum at the *speed of light*:

$$c_0 = 299\,792.458 \, \frac{\text{km}}{\text{s}}$$

The rounded value of

$$c_0 = 300\,000 \, \frac{\text{km}}{\text{s}} = 3 \cdot 10^5 \, \frac{\text{km}}{\text{s}} = 3 \cdot 10^8 \, \frac{\text{m}}{\text{s}}$$

is sufficiently accurate to describe the propagation of light in air.

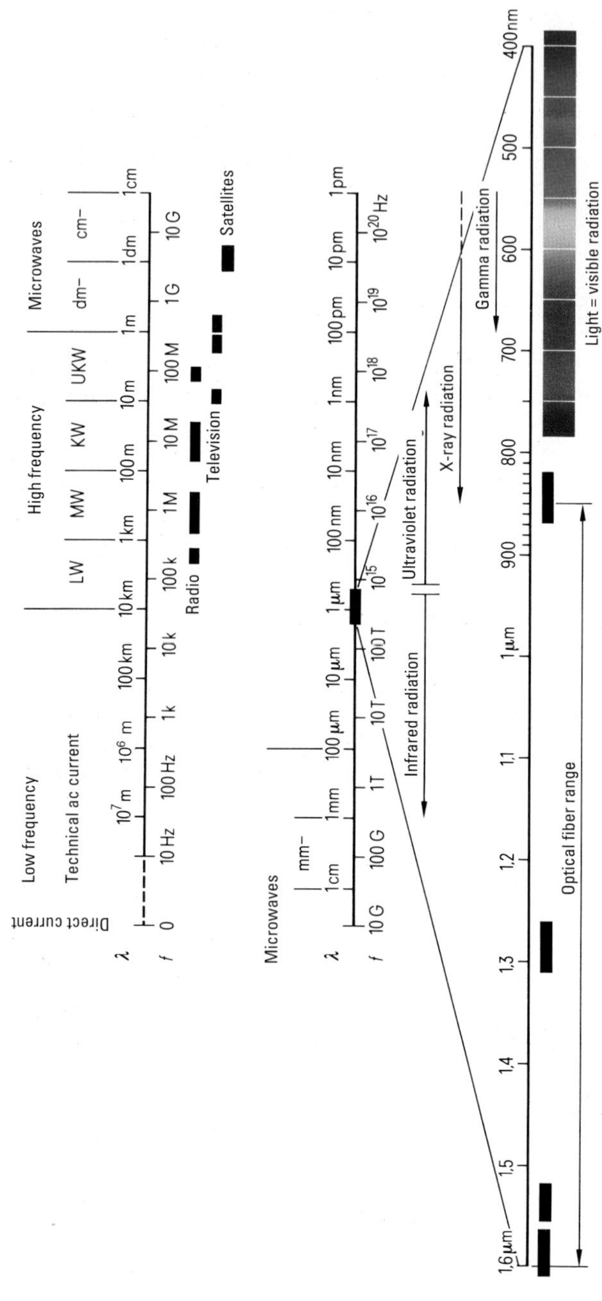

Figure 2.1 Spectrum of the electromagnetic waves

In a nondissipative, infinite medium, the electromagnetic wave and therefore the light wave, too, is a transverse wave. Its electric and magnetic field oscillates perpendicular to the direction of propagation. If the electric or magnetic field oscillates in a plane, then the tip of the pointer of the electric or magnetic field follows a straight line; such a wave is said to be linearly polarized. If the tip of the pointer describes a circle or more generally an ellipse, then the terms circular or elliptical polarization are used. Figure 2.2 shows the different types of polarization for a light wave propagating in the z-direction.

2.2 Fundamental Concepts of Wave Theory

In general, a wave is understood to be the propagation of a condition or its generation in a medium without actually transporting the mass or material of the medium itself. In the case of a light wave, the condition is an electromagnetic field which propagates in a transparent material, the optical medium. In the simplest case, the propagation of such a wave in time and space can be described by a sine function. Therefore with regard to the displacement a of a plane wave propagating in the z direction, the following applies:

$$a = A \cdot \sin(\omega t - kz) = A \cdot \sin 2\pi \left(\frac{t}{T} - \frac{z}{\lambda} \right).$$

- a displacement of a plane wave (e.g. magnetic or electric field strength or its squares (intensity))
- A amplitude in units of the displacement
- ω angular frequency in s^{-1}
- t time in s
- k wave number in m^{-1}
- z length in z direction in m
- T period of oscillation in s
- λ wavelength in m.

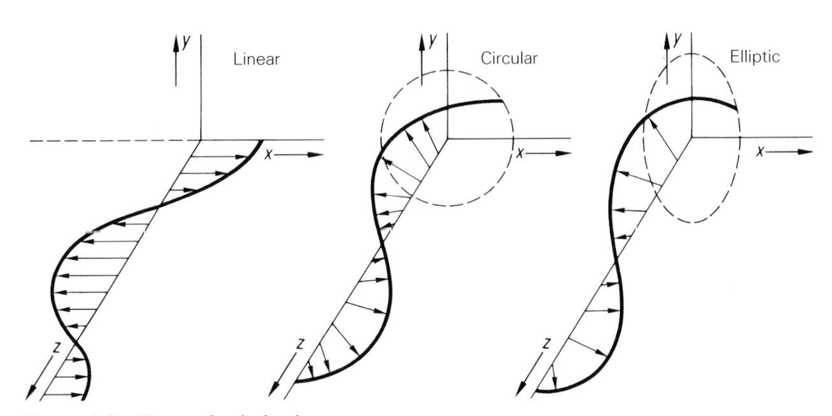

Figure 2.2 Types of polarization

The amplitude A of the wave describes the largest displacement from the equilibrium condition. The expression in parentheses $(\omega t - kz)$ is called the *phase angle* or, abbreviated, the phase of the wave. The phase angle φ is a circular measure expressed in rad (radians), e.g.

$$1 \text{ rad} = \frac{360°}{2\pi} = 57.295°.$$

For further clarification, Figure 2.3 shows a plane wave as a function of the time t at a fixed point $z = z_0$. It can be seen that the oscillating points a_1 and a_3 are in the same phase of oscillation. Their phase difference is 2π. By contrast, the point a_2 has the same displacement; however, it is in a different phase.

The term ω is called the *angular frequency*. It is equal to 2π times the frequency f, which is used to express the number of oscillations per unit of time. Therefore it holds:

$$\omega = 2\pi \cdot f$$

where f is the frequency in Hz (hertz) or cps (cycles per second). Some derived units for the frequency can be taken from Table 2.1.

T is used to describe the *period* of oscillation, i.e. the time of one full oscillation or one complete cycle. Some derived units for the period of oscillation are given in Table 2.2. The frequency f is inversely proportional to the period of oscillation T:

$$f = \frac{1}{T}.$$

where T is the period of oscillation in s.

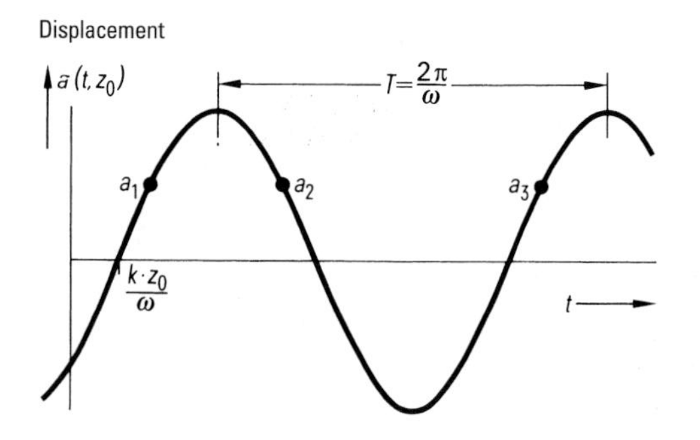

Figure 2.3
Plane harmonic wave at a fixed point z_0

The symbol k is used to describe the *wave number*. It is equal to the magnitude of the wave vector $\vec{k}$ which indicates the direction of propagation of the wave. The wave number k gives the phase shift of the wave per unit of the length and is therefore inversely proportional to the wavelength λ with the proportionality factor 2π:

$$k = \frac{2\pi}{\lambda}$$

The wavelength λ is the spatial period of the wave, i.e. the distance or length of path of a complete oscillation. Units derived from m can be seen in Table 2.3.

The following basic relation exists between the frequency f, the wavelength λ and the speed of propagation c of a wave:

$$c = f \cdot \lambda$$

Table 2.1
Units derived from Hz

Unit	Name
1 kHz $= 10^3$ Hz	Kilohertz
1 MHz $= 10^6$ Hz	Megahertz
1 GHz $= 10^9$ Hz	Gigahertz
1 THz $= 10^{12}$ Hz	Terahertz

Table 2.2
Units derived from s

Unit	Name
1 ms $= 10^{-3}$ s	Millisecond
1 µs $= 10^{-6}$ s	Microsecond
1 ns $= 10^{-9}$ s	Nanosecond
1 ps $= 10^{-12}$ s	Picosecond
1 fs $= 10^{-15}$ s	Femtosecond

Table 2.3
Units derived from m

Unit	Name
1 Mm $= 10^6$ m	Megameter
1 km $= 10^3$ m	Kilometer
1 m $= 1$ m	Meter
1 dm $= 10^{-1}$ m	Decimeter
1 cm $= 10^{-2}$ m	Centimeter
1 mm $= 10^{-3}$ m	Millimeter
1 µm $= 10^{-6}$ m	Micrometer
1 nm $= 10^{-9}$ m	Nanometer

Example

Light with a wavelength λ = 1 µm has a speed of propagation in air of c_0 = 300 000 $\frac{\text{km}}{\text{s}}$. The frequency f of the light wave is then:

$$f = \frac{c_0}{\lambda} = \frac{300\,000\,\frac{\text{km}}{\text{s}}}{1\,\mu\text{m}} = \frac{3 \cdot 10^8\,\frac{\text{m}}{\text{s}}}{1 \cdot 10^{-6}\,\text{m}};$$

$$f = 3 \cdot 10^{14}\,\frac{1}{\text{s}} = 300 \cdot 10^{12}\ \text{Hz} = 300\ \text{THz}.$$

2.3 Reflection of Light

When light falls on the interface between two media, a percentage of it is reflected.

The amount of light reflected depends on the angle α_1 between the incident ray of light and the normal of incidence. Ray of light is used here to describe the path in which light energy travels. For the reflected ray and the angle α_2, created by the normal of incidence and the reflected light ray, the following applies (Figure 2.4):

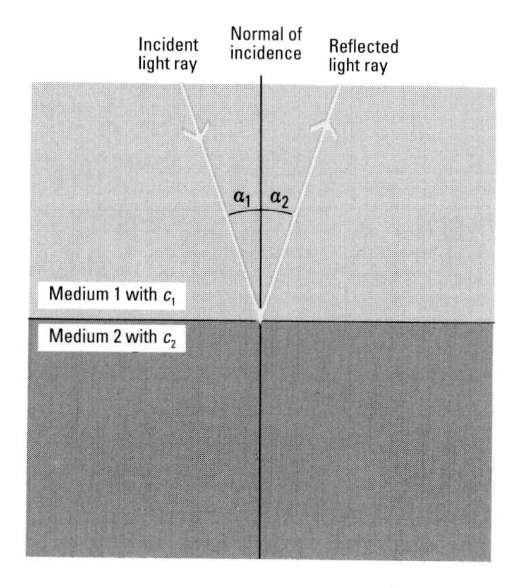

Figure 2.4
Reflection of light

The reflected ray

- ▷ remains in the plane of incidence described by the incident light ray and the normal of incidence,
- ▷ lies relative to the incident ray on the opposite side of the normal of incidence, and
- ▷ is at the same angle to the normal of incidence:

$$\alpha_1 = \alpha_2$$

2.4 Refraction of Light

When a ray of light with an angle of incidence α enters an optically denser medium (e.g. glass or water) from an optically less dense one (e.g. air), its direction is bent toward an angle of refraction β.

In the case of an isotropic medium, i.e. a material or substance that has identical properties in all directions, *Snell's law of refraction* applies:

The ratio of the sine of the *angle of incidence* α to the sine of the *angle of refraction* β is constant and also identical to the ratio c_1/c_2 of the speeds of light c_1 in the first and c_2 in the second medium (Figure 2.5):

$$\frac{\sin \alpha}{\sin \beta} = \frac{c_1}{c_2}.$$

α	angle of incidence	c_1	speed of light 1
β	angle of refraction	c_2	speed of light 2.

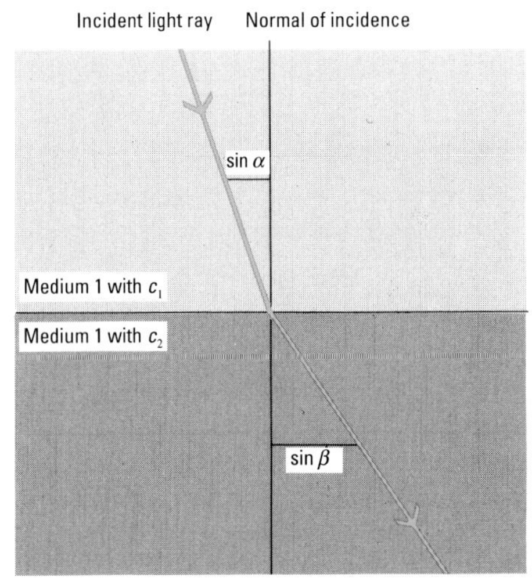

Figure 2.5
Refraction of light

With two transparent media, the one in which the speed of light is slower is described as denser.

For the transition from a vacuum (≈ air) in which light travels at a speed c_0 to a medium with a speed of light c the following applies:

$$\frac{\sin\alpha}{\sin\beta} = \frac{c_0}{c} = n.$$

The ratio of the speed of light c_0 in vacuum to the speed of light c in the medium is called the *refractive index n* (more precisely the *phase refractive index*) of the respective medium. The refractive index of vacuum (≈ air) n_0 is equal to 1.

For two different media with the refractive indices n_1 and n_2 and their speeds of light c_1 and c_2 the following applies:

$$c_1 = \frac{c_0}{n_1} \text{ and } c_2 = \frac{c_0}{n_2}.$$

Another form of Snell's law of refraction is derived from that:

$$\frac{\sin\alpha}{\sin\beta} = \frac{n_2}{n_1}.$$

The sine of the angle of incidence relative to the sine of the angle of refraction is inversely proportional to their refractive indices.

Example

At a refractive index of $n_1 = 1.5$, which is generally assumed for the glass in an optical fiber, the speed of light c_1 in the fiber is

$$c_1 = \frac{c_0}{n} = \frac{300\,000\,\frac{\text{km}}{\text{s}}}{1.5};$$

$$c_1 = 200\,000\,\frac{\text{km}}{\text{s}} = 200\,\frac{\text{m}}{\mu\text{s}}$$

or 5 μs for each km of fiber or
5 ns for each m of fiber.

The refractive index n of a medium depends fundamentally on the wavelength of the light; for infrared wavelengths which are important for the optical communication in fused silica glass it decreases constantly as the wavelength increases.

The quantity n applies for light waves which propagate in only one wavelength and with a constant amplitude and therefore cannot transmit any information. Only *modulation* of these waves makes it possible to transport information. In (digital) optical communication, light pulses are used for this purposes. These are short wave groups which contain light waves of different wavelengths.

Within such wave groups individual waves propagate at different speeds due to their different wavelengths. The speed of propagation of such a wave group is called the *group velocity*. The associated group refractive index n_g has been defined and is related to the refractive index as follows:

$$n_g = n - \lambda \frac{dn}{d\lambda} .$$

The curves of n and n_g for pure fused silica glass as a function of the wavelength λ are shown in Figure 2.6. Some numerical values may be seen in Table 2.4.

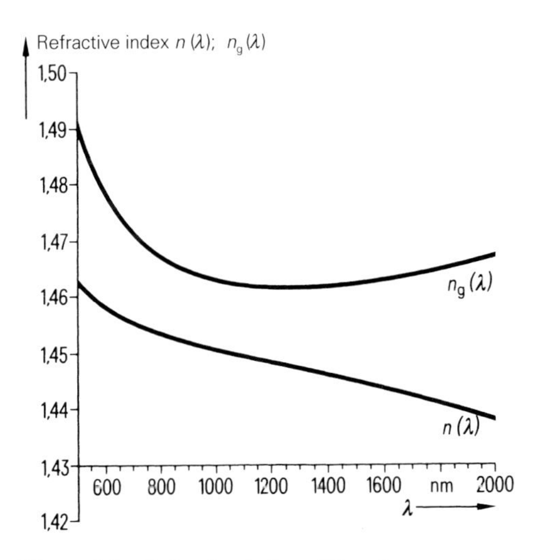

Figure 2.6 Refractive index $n(\lambda)$ and group refractive index $n_g(\lambda)$ (100% SiO_2)

Table 2.4
Refractive index n (λ) and group refractive index n_g (λ) (100 % SiO_2)

Wavelength (λ) nm	Refractive index n	Group refractive index n_g
600	1.4580	1.4780
700	1.4553	1.4712
800	1.4533	1.4671
900	1.4518	1.4646
1000	1.4504	1.4630
1100	1.4492	1.4621
1200	1.4481	1.4617
1300	1.4469	1.4616
1400	1.4458	1.4618
1500	1.4446	1.4623
1600	1.4434	1.4629
1700	1.4422	1.4638
1800	1.4409	1.4648

The expression $\frac{dn}{d\lambda}$ gives the slope of the refractive index curve $n(\lambda)$, which is downward (negative) in the observed wavelength range. Therefore the group refractive index n_g at each wavelength is larger than the refractive index n. For calculations of transmission times of optical signals only the group refractive index n_g is to be used.

It is useful to note that the group refractive index reaches a minimum near the wavelength of 1300 nm. This wavelength range is particularly interesting for optical transmission.

2.5 Total Internal Reflection

When a light ray contacts the interface between an optically dense medium with a refractive index n_1 and an optically less dense medium with a refractive index n_2 at an ever smaller angle, i.e. at an ever higher angle of incidence α, then at a certain angle of incidence α_0 the angle of refraction becomes β_0 = 90° (Figure 2.7).

In this case the light ray (2) propagates parallel to the interface of the two media. The angle of incidence α_0 is called the *critical angle* of the two media.

For the critical angle α_0 the following ratio applies:

$$\sin \alpha_0 = \frac{n_2}{n_1},$$

i.e. the critical angle is dependent on the ratio of the refractive indices n_1 and n_2 of the two media.

Examples

The critical angle between water with ($n_1 = 1.333$) and air with ($n_0 = 1$) is:

$$\sin \alpha_0 = \frac{1}{1.333} \approx 0.75 \text{ and } \alpha_0 \approx 49°$$

between glass with ($n_1 = 1.5$) and air with ($n_0 = 1$) it is

$$\sin \alpha_0 = \frac{1}{1.5} \approx 0.67 \text{ and } \alpha_0 \approx 42°.$$

For all rays with an angle of incidence α greater than the critical angle α_0 there are no corresponding refracted rays in the optically thinner medium. These light rays are reflected at the interface back into the denser medium.

This phenomenon is called *total internal reflection* (light ray 1).

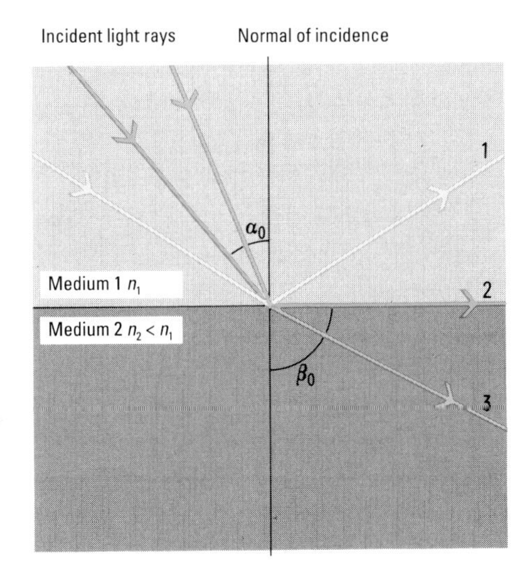

1 Totally reflected light ray
2 Refracted light ray with angle of refraction $\beta = 90°$
3 Refracted light ray

Figure 2.7
Total internal reflection of light

Total internal reflection can only occur at an interface where a light ray propagates from an optically denser medium (e.g. glass $n_1 = 1.5$) into an optically less dense medium (e.g. air $n_0 = 1$), never in the reverse case.

2.6 Numerical Aperture

The effect of total internal reflection is used in optical waveguides by having "core glass" in the middle of the waveguide with a refractive index of n_1 and around it "cladding glass" with a refractive index n_2 with n_1 being slightly higher than n_2 (Figure 2.8).

From the requirement $\sin \alpha_0 = \frac{n_2}{n_1}$ it follows that all rays that are not divergent to the fiber axis by more than the angle $(90° - \alpha_0)$ will be guided in the core glass.

In order to launch light from outside (air with refractive index $n_0 = 1$) into the core glass, the launch angle between the light ray and the fiber axis can be determined according to the law of refraction:

$$\frac{\sin \Theta}{\sin (90° - \alpha_0)} = \frac{n_1}{n_0}$$

and hence:

$$\sin \Theta = n_1 \cdot \cos \alpha_0 = n_1 \cdot \sqrt{1 - \sin^2 \alpha_0} \ .$$

Together with the requirement for the critical angle $\alpha_0 = \frac{n_2}{n_1}$ the result is

$$\sin \Theta = \sqrt{n_1^2 - n_2^2} \ .$$

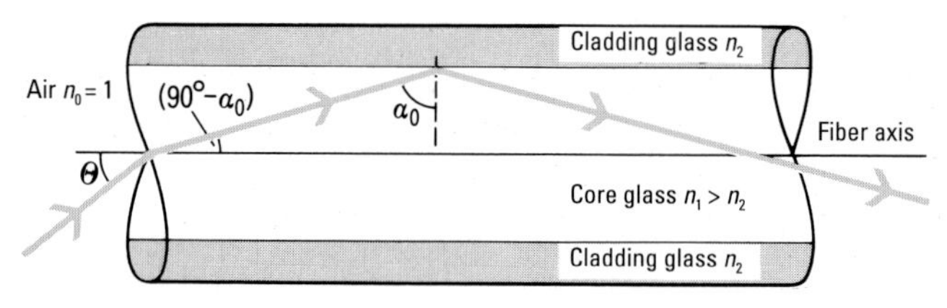

Figure 2.8 Light guidance in an optical waveguide

The greatest possible launch angle Θ_{max} is called the *acceptance angle* of the optical fiber; it is only dependent on the two refractive indices n_1 and n_2. The sine of the acceptance angle is called the *numerical aperture NA* of the optical fiber:

$$NA = \sin \Theta_{max}$$

This quantity has major importance in launching light into the optical fiber.

2.7 Propagation of Light in an Optical Waveguide

Optical laws make it possible to describe the total internal reflection of light at the interface between core glass and cladding glass of an optical fiber. A ray-shaped linear propagation of the light is the basic assumption. In order to more closely study the various possibilities of the propagation of light inside the core glass, it is necessary to take into consideration phenomena typical of wave optics. In particular this proves necessary when the diameter of the fiber core is about 10 μm and therefore only somewhat larger than the wavelength of the light guided in the core of about 1 μm. Thus interference patterns occur that can be explained with the help of wave optics.

The superposition of two or more waves and their merger into a single wave is generally called *interference*. A clear pattern of interference of two waves can only be obtained when both waves are of the same wavelength and have a constant phase difference at any time to each other. Such waves are called *coherent waves*. If the two waves differ in phase at a certain point in space by an integer multiple of their wavelength λ then *addition* of the amplitude occurs. On the other hand, in the case of a phase difference of a multiple of half of the wavelength $\lambda/2$ *subtraction* occurs and in the case of two waves with identical amplitude the waves are *cancelled out* at that point (destructive interference).

When two normal sources of light are used, e.g. light bulbs, and their light is superimposed, then no interference can be observed, because this light is incoherent. This is due to the way light is radiated by the incandescent body, in this case the filament in the bulb.

Due to spontaneous processes at random times the individual atoms of the filament emit light flashes, i.e. short wave groups of about 10^{-8} s duration. At the speed of light in air of 3×10^8 m/s these wave groups are about 3 m long. This length is described as *coherence length*. The superposition of the 3 m long wave groups is completely random and merely brightens the room.

In order to transmit light in optical waveguides, it was necessary to find a source of highly coherent light. The spectral width of the transmitter (Chapter 12.1) should have been as small as possible. In contrast to light emitting diodes with a spectral width of ≥ 40 nm, lasers with their stimulated light emission make it possible to maintain a constant phase difference at one wavelength. Therefore interference phenomena occur in an optical waveguide, which can be recognized by the fact that light propagates in the core glass only under precise angles; "precise" means that it must propagate in directions in which the light waves involved are amplified due to their superposition and therefore show constructive interference. The permissible light waves which can propagate in an optical fiber are called *modes* (eigenwaves).

These modes can be determined with mathematical precision by means of *Maxwell's equations*. This general system of equations for electromagnetic waves can be greatly simplified when applied to light waves which are only weakly guided in the optical fiber. This applies for waves which propagate almost exactly in the direction of the fiber axis and whose field components along the axis are negligible. They occur when the refractive index n_1 of the core glass and n_2 of the cladding glass differ only slightly. A measure for this difference in refraction is the normalized *refractive index difference* Δ. It is defined as

$$\Delta = \frac{n_1^2 - n_2^2}{2n_1^2} \approx \frac{n_1 - n_2}{n_1} .$$

In an optical fiber the normalized refractive index difference Δ is very small compared to unity and the light waves are therefore guided correspondingly weakly in the core glass.

Solutions of the simplified wave equations offer very good approximations for the propagative modes in an optical fiber. As an example Figure 2.9 shows the distribution of light intensity over the cross-section of an optical fiber for the first ten modes. The eigenwaves have plane wave surfaces and are *l*inearly *p*olarized. Therefore they are termed $LP_{\nu\mu}$ with the two mode numbers ν and μ.

The quantity ν is the azimuthal mode number. It is used to give half the number of light points in each concentric light ring. The values for ν can be 0, 1, 2, 3, ..., whereby at $\nu = 0$ each light ring appears without subdivisions.

The quantity μ is the radial mode number; it is used to give the number of concentric light rings of the mode. It can take the values 1, 2, 3, ...

The fundamental mode is called LP_{01}; the next higher mode, LP_{11}.

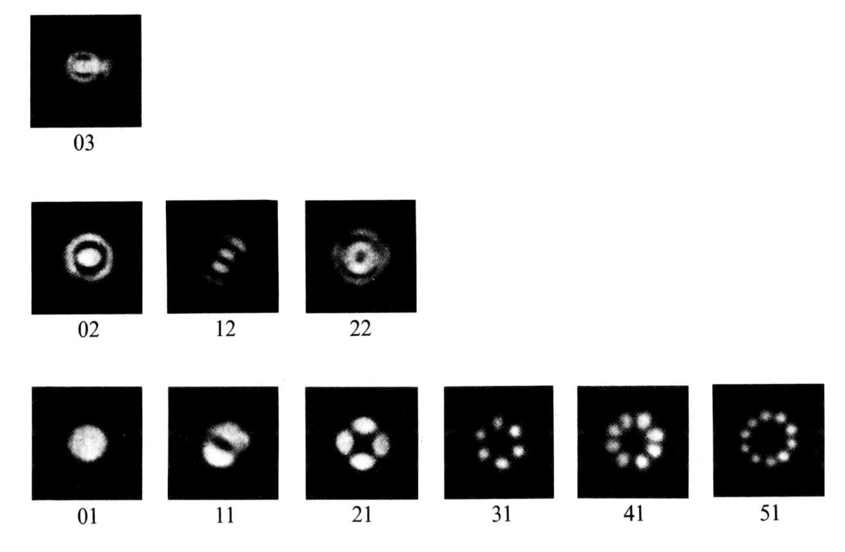

Figure 2.9 The first ten modes $LP_{\nu\mu}$ of an optical waveguide
(Stolen, R.H., and Leibolt, W.N.: Optical fiber modes using stimulated four-photon mixing. *Appl. Opt.*, **15,** 1976, 239-243)

2.8 Optical Nonlinear Effects

When an optical fiber operates with light at high intensity, i.e. high light power per fiber cross-sectional area, optical nonlinear effects become significant. These effects are the consequence of the interaction of incident photons with the glass material of the fiber at the subatomic level. Hence for a deep understanding of the physics involved one must deal with quantum mechanics.

Nonlinearity in an optical fiber is not a design or manufacturing defect. It is rather an inherent characteristic of light energy passing through a fiber and, therefore, of particular concern for the design engineers of fiber-based communication systems.

The effects of optical nonlinearity are divided into scattering phenomena and refractive index phenomena. They can loosely be grouped into at least four types. The first type is the interaction of incident photons with molecular vibrations (optical phonons), known as *stimulated Raman scattering*. The second type occurs when photons interact with acoustic phonons (sound vibrations), known as *stimulated Brillouin scattering (SBS)*.

Both scattering mechanisms result in a shift of photon wavelength and are referred to as inelastic scattering, because part of the incident light energy is transferred to the optical fiber material The third type is caused by the intensity-dependent changes of the refractive index when photons interact with lattice electrons instantaneously. In isotropic SiO_2 glass this effect leads to processes, known as *4-photon mixing*, also called *four-wave mixing (FWM),* as well as *self-phase modulation (SPM)* and *cross-phase modulation.* The fourth type involves the trapping of a perturbed electron, so that the refractive index change is long lasting. This effect is known as *photorefractive effect.*

Nonlinear effects in optical fibers have gained a significant influence on the operation of long-haul optical communication systems. In particular, the use of dense wavelength division multiplex (DWDM) systems (Chapter 13), high power light sources, smaller spectral width lasers and small fiber cores lead to power densities which are high enough to exceed the critical threshold level for nonlinear effects like four-wave mixing, stimulated Brillouin scattering and self-phase modulation.

The threshold power levels are in the order of 190 mW for the SBS and approximately 30 mW for the SPM respectively. The other above mentioned nonlinear effects like stimulated Raman scattering require higher power densities and thus are not as harmful in practical applications. FWM, SBS and SPM can lead to severe signal distortions.

Four-wave mixing can be one of the most disruptive nonlinear effects in wavelength division multiplex (WDM) systems. When the intensity of the laser signal reaches a critical value, ghost channels appear, some of which may fall within the true channels. The net result of FWM is that several wavelengths mix to produce new wavelengths. The number of new wavelengths (or ghosts) is calculated as $N^2(N\text{-}1)/2$, where N is the number of original wavelengths. Thus, 24 ghost channels appear in a 4-channel system (N=4). With N=16 there will be 1920 ghost channels. Interference of this magnitude can be catastrophic at the receiver end.

In the SBS process a significant proportion of the optical power travelling through the fiber may be converted into a second light wave, shifted in frequency, travelling backwards towards the transmitter. SBS can be detrimental in an optical communications system in a number of ways by introducing severe additional signal attenuation, by causing multiple frequency shifts in some cases and by introducing a high intensity backwards coupling into the transmission optics. It will therefore be necessary to operate at power levels below the threshold for SBS, and this could place a severe limitation on the launch power and thus also the repeater spacing.

Self-phase modulation leads to an intensity dependent broadening of the signal spectrum. This greater frequency width increases pulse spreading through group velocity dispersion. The process can be understood as a differential phase shift between the center and the tails of the optical pulse due to the intensity dependent refractive index. The index change is extremely small but the corresponding phase modulation can be significant when added up in a long fiber.

If, however, SPM and chromatic dispersion are well adjusted to each other (correct light intensity, negative dispersion) pulse narrowing instead of pulse broadening can occur, which can lead to optical *solitons*. Solitons are light pulses that are not subject to deformation by dispersion in optical fibers and therefore in principle do not experience any limitation in bit rates over long transmission lengths. Thus they seem to be particularly well suited for telecommunication especially along transoceanic transmission lines. Soliton transmission has been demonstrated successfully in laboratories up to a fiber length of 9000 km. Commercially used soliton based transmission lines are not installed yet.

A key factor in nonlinear effects is the *nonlinear coefficient* γ, which is defined as:

$$\gamma = \frac{2\pi}{\lambda} \cdot \frac{n_2}{A_{eff}}$$

λ	wavelength
n_2	nonlinear refractive index
A_{eff}	effective area

The *effective area* A_{eff} is the area of a fiber over which an assumed constant intensity will produce the same nonlinear effects as the actual varying intensity. A small effective area means higher laser signal intensity, which leads to stronger nonlinear effects. In practice, A_{eff} is proportional to the fiber core area. If one assumes that the light beam propagates in a Gaussian form, then

$A_{eff} = \pi\, w_0^2 /4,$

where w_0 is the mode field radius (Chapter 5.6).

Nonlinear effects increase with the length of the fiber span. As the core area is normalized with respect to nonlinear effects using A_{eff}, the fiber span is normalized with respect to nonlinear effects using the *effective fiber length* L_{eff}. The definition of L_{eff} is analogous to A_{eff}. L_{eff} is typically equal to 20 km.

In order to counterbalance or at least reduce nonlinear effects a specially designed single-mode optical fiber, called nonzero dispersion shifted fiber

(NZDSF), has been developed (Chapter 6.1). It has been designed to overcome the pulse broadening and FWM impairments that can occur in optically amplified (EDFA) multiple wavelengths (DWDM) systems (Chapter 13) operating at high bit rates.

The NZDSF fiber accomplishes this by using a large effective area and a small but finite amount of chromatic dispersion over the EDFA wavelength range. The dispersion is low enough to carry high bit rate pulses over long distances without requiring dispersion compensating equipment. At the same time, the fiber's dispersion is high enough to suppress the FWM nonlinearity by reducing phase matching between the wavelengths used in DWDM systems. By increasing the effective area of the mode field within the fiber the overall power density, and thereby, the nonlinear effects, are being reduced as well. This benefit leads to higher power-handling capability, greater signal-to-noise ratio, and greater amplifier spacing.

3 Chemistry of Optical Waveguides

3.1 Fused Silica Glass

In terms of weight the outer solid crust of the earth is composed of one-half oxygen and one-fourth silicon. In the periodic table of the elements oxygen is element 8 and silicon is element 14 (Table 3.1).

The great abundance of those two elements is due to the fact that the earth's crust is composed mainly of quartz and its compounds with metallic oxides, the silicates. Quartz, which is called *silicon dioxide* SiO_2 as a chemical compound, appears largely as quartzite, one of the components of sand. Weathering of igneous rock, mainly granite, has created this sand in the course of the history of the earth.

In its purest form crystalline quartz can be found as rock crystal, which is as clear as water. Its optical and mechanical properties are anisotropic, i.e. they differ with the direction of the individual axes of the crystal.

Because of its manifold applications in technology, e.g. as quartz resonator, as an optically active component or in piezoelectricity, quartz is produced synthetically today by stimulating crystal growth around a nucleus.

In contrast to quartz, fused glass is an amorphous, i.e. noncrystalline, glassy solidified melt of silicon dioxide which appears only to be a solid due to its high viscosity. It has no melting point; on the contrary, at higher temperatures it merely becomes increasingly soft and vaporizes directly from this state without going through a liquid state.

The viscosity is an essential property for the entire production and forming process of glass. With it the inner friction in fused silica glass is described. It is named η and can be given in units of decipascalseconds, whereby

$$(\text{dPa} \cdot \text{s}) = \frac{1\,\text{g}}{\text{cm} \cdot \text{s}}$$

In fused silica glass the viscosity decreases monotonically with the temperature T (Figure 3.1).

Table 3.1 Periodic table of the elements

Key

Atomic number	1 H Hydrogen 1.00.797	Symbol
	Atomic weight (or mass number of most stable isotope if in parentheses)	

1a	2a	3b	4b	5b	6b	7b	8			1b	2b	3a	4a	5a	6a	7a	0
1 **H** Hydrogen 1.00.797																	2 **He** Helium 4.0026
3 **Li** Lithium 6.939	4 **Be** Beryllium 9.0122											5 **B** Boron 10.811	6 **C** Carbon 12.01115	7 **N** Nitrogen 14.0067	8 **O** Oxygen 15.09984	9 **F** Fluorine 18.9984	10 **Ne** Neon 20.179
11 **Na** Sodium 22.9898	12 **Mg** Magnesium 24.305											13 **Al** Aluminium 26.9815	14 **Si** Silicon 28.086	15 **P** Phosphorus 30.9738	16 **S** Sulfur 32.064	17 **Cl** Chlorine 35.453	18 **Ar** Argon 39.948
19 **K** Potassium 39.102	20 **Ca** Calcium 40.08	21 **Sc** Scandium 44.956	22 **Ti** Titanium 47.90	23 **V** Vanadium 50.942	24 **Cr** Chromium 51.6996	25 **Mn** Manganese 54.9380	26 **Fe** Iron 55.847	27 **Co** Cobalt 58.9332	28 **Ni** Nickel 58.71	29 **Cu** Copper 63.546	30 **Zn** Zinc 65.37	31 **Ga** Gallium 69.72	32 **Ge** Germanium 72.59	33 **As** Arsenic 74.9216	34 **Se** Selenium 78.96	35 **Br** Bromine 79.904	36 **Kr** Krypton 83.80
37 **Rb** Rubidium 85.47	38 **Sr** Strontium 87.62	39 **Y** Yttium 89.905	40 **Zr** Zirconium 91.22	41 **Nb** Niobium 92.906	42 **Mo** Molybdenum 95.94	43 **Tc** Technetium (97)	44 Ru Ruthenium 1001.07	45 **Rh** Rhodium 102.905	46 **Pd** Palladium 106.4	47 **Ag** Silver 107.868	48 **Cd** Cadmuim 112.40	49 **In** Indium 114.82	50 **Sn** Tin 118.69	51 **Sb** Antimony 121.75	52 **Te** Tellurium 127.60	53 **I** Iodine 126.9044	54 **Xe** Xenon 131.30
55 **Cs** Cesium 132.905	56 **Ba** Barium 137.34	57-71* *Lanthanides	72 **Hf** Hafium 178.49	73 **Ta** Tantalum 180.948	74 **W** Tungsten 183.85	75 **Re** Rhenium 186.2	76 **Os** Osmium 190.2	77 **Ir** Iridium 192.2	78 **Pt** Platinum 195.09	79 **Au** Gold 196.967	80 **Hg** Mercury 200.59	81 **Tl** Thallium 204.37	82 **Pb** Lead 207.19	83 **Bi** Bismuth 208.980	84 **Po** Polonium (210)	85 **At** Astatine (210)	86 **Rn** Radon (222)
87 **Fr** Francium (223)	88 **Ra** Radium (226)	89-103** **Actinides															

*Lanthanides	57 **La** Lanthanum 138.91	58 **Ce** Cerium 140.12	59 **Pr** Praseodymium 140.907	60 **Nd** Neodymium 144.24	61 **Pm** Promethium (145)	62 **Sm** Samarium 150.35	63 **Eu** Europium 151.96	64 **Gd** Gadolinium 157.25	65 **Tb** Terbium 158.924	66 **Dy** Dysprosium 162.50	67 **Ho** Holmium 164.930	68 **Er** Erbium 167.26	69 **Tm** Thulium 168.934	70 **Yb** Ytterbium 173.04	71 **Lu** Lutetium 174.97
**Actinides	89 **Ac** Actinium (227)	90 **Th** Thorium 232.038	91 **Pa** Protactinium (231)	92 **U** Uranium 238.03	93 **Np** Neptunium (237)	94 **Pu** Plutonium (244)	95 **Am** Americium (243)	96 **Cm** Curium (247)	97 **Bk** Berkelium (247)	98 **Cf** Californium (251)	99 **Es** Einsteinium (254)	100 **Fm** Fermium (257)	101 **Md** Mendelevium (257)	102 **No** Nobelium (255)	103 **Lr** Lawrencium (256)

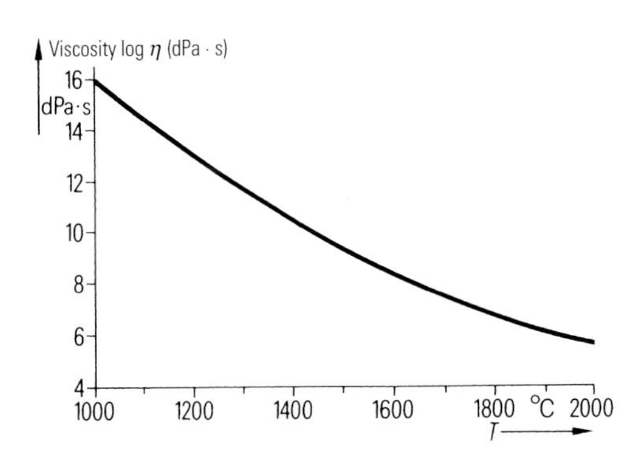

Figure 3.1
Viscosity of fused silica glass as a function of temperature

For practical reasons several temperatures have been more closely defined with the aid of the logarithm of the viscosity (Table 3.2), because the viscosity-temperature curve has no prominent points.

The upper and lower annealing points are at the boundaries of the transformation range, i.e. at the transition point from the viscoelastic condition of a melt to the brittle state of fused silica glass. Within the range of the softening point the shape of the fused silica glass body changes due to its own weight.

3.1.1 Production

Ultra pure fused silica glass is usually produced by deposition of SiO_2 out of the vapor phase by transforming the highly volatile silicon tetrachloride compound with oxygen giving off chlorine gas. The detour via $SiCl_4$ is chosen because in contrast to natural SiO_2 this compound can be produced in a very pure form by distillation. The reaction equation is

$$SiCl_4 + O_2 \xrightarrow{1700°C} SiCO_2 + 2\,Cl_2.$$

Table 3.2
Description of several temperature values for fused silica glass by means of log η

Viscosity log η	Description	Temperature in fused silica glass °C
7.6	Softening point	1730
13	Annealing point	1180
14.5	Strain point	1075

Optical waveguides are produced today according to this process for use in optical communications. It is well known that the refractive index of the glass is an important factor for the propagation of light within the optical waveguide. It can be "adjusted" by appropriate doping during the vapor deposition process, i.e. by adding precise amounts of oxides. For example by incorporating fluorine (F) or boron trioxide (B_2O_3) a lower refractive index can be obtained and by adding germanium dioxide (GeO_2) or phosphorus pentaoxide (P_2O_5) a higher refractive index is obtained, such as is needed for the core glass of the optical fiber. However, the refractive index differences which can be created by these means in fused silica glass are relatively limited.

In Figure 3.2 the refractive index *n* of doped fused silica glass is shown as a function of the dopant concentration of these materials.

By incorporating these oxides in the ultra pure silicon dioxide not only the refractive index changes; other factors are also affected. Thus the linear expansion characteristics due to temperature changes of doped and undoped fused silica glass are different. It is also particularly important that by incorporation of foreign molecules the light scattering and thereby the attenuation of the propagating light is increased.

A further cause of attenuation as light travels through fused silica glass is *absorption* by the transition metals, Fe, Cu, Co, Cr, Ni, Mn and by water in the

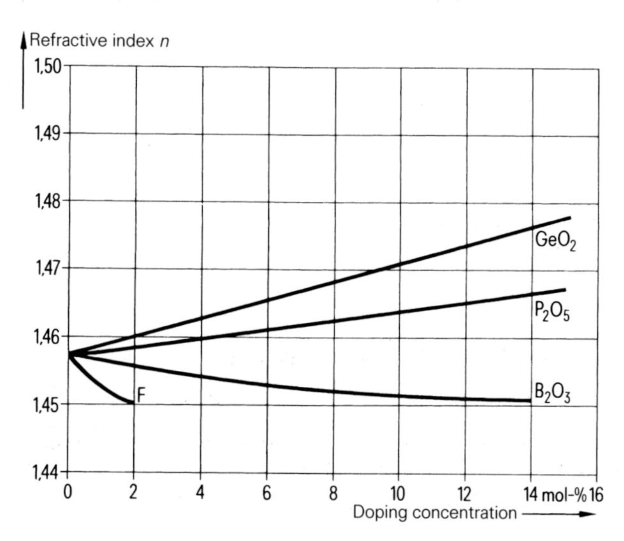

Figure 3.2 Refractive index of SiO_2 with different dopant materials

form of OH ions. At very minute contaminations of the glass with these metals and OH ions high light losses occur. The concentration of these impurities is measured in ppm (parts per million) or ppb (parts per billion), i.e. one part of the contaminant per million or billion of the base substance. For example, 1 ppm Cu causes several hundred decibels per kilometer attenuation in the 800 nm range and an OH concentration of 1 ppm causes an attenuation of 0.1 dB/km at 800 nm, 1 dB/km at 950 nm, 1.7 dB/km at 1240 nm and 35 dB/km at 1390 nm. Hence it is clear that, dependent on the kind of impurity, the absorption is particularly pronounced at certain wavelengths, the so-called absorption bands.

When *multicomponent glass* such as lead alkali silicate glass or sodium borosilicate glass is used as the raw material for optical waveguide production instead of ultra pure silicon dioxide, then higher attenuation results due to the impurities. Normal glass, e.g. for window panes or drinking glasses, contains additional oxides and therefore it is much less translucent; on the other hand, it has mechanical and production advantages.

3.1.2 Material Properties

Fused silica glass is an isotropic medium; i.e. independent of direction its physical properties are identical. Its behavior is well known when temperatures change quickly. Due to its extremely small *linear thermal expansion coefficient* α (Table 3.3), it is extraordinarily resistant to temperature changes.

Example

Change in the length of a fused silica glass fiber:

Test length $L = 1$ km, temperature change from 20 to 40 °C:

$$\Delta T = 20 \text{ K}.$$

The change in length ΔL is calculated:

$$\Delta L = \alpha \cdot \Delta T \cdot L;$$
$$\Delta L = 5.5 \cdot 10^{-7}/\text{K} \cdot 20 \text{ K} \cdot 1 \text{ km} = 110 \cdot 10^{-7} \cdot 10^{5} \text{ cm};$$
$$\Delta L = 1.1 \text{ cm}.$$

Some further typical properties of fused silica glass an be found in Table 3.3.

The following sample calculations will clarify these properties:

▷ *Calculation of the weight* of a fused silica glass fiber without coating with a diameter ($d = 125$ µm = 0.125 mm), 1 km in length.

The surface of the cross-section of the glass is

$$A = \pi \cdot \frac{d^2}{4} \approx 3.14 \cdot \frac{0{,}125^2 \text{ mm}^2}{4} \approx 0.0123 \text{ mm}^2 = 1.23 \cdot 10^{-4} \text{ cm}^2.$$

The weight G per kilometer then results as

$$G = \gamma \cdot A \cdot 1\ \text{km} = 2.20\ \frac{\text{g}}{\text{cm}^3} \cdot 1.23 \cdot 10^{-4}\ \text{cm}^2 \cdot 10^5\ \text{cm} \approx 27\ \text{g}.$$

▷ *Calculation of the longitudinal strain* ε and the corresponding tensile force of a fused silica glass fiber with a diameter d of 125 µm = 0.125 mm in a proof test with a tension of.

$$\sigma \approx 345\ \frac{\text{N}}{\text{mm}^2} \approx 50\ \text{KPSI}^{1)}.$$

Hooke's law states:

$$\sigma = E \cdot \varepsilon.$$

The stress σ (tension or compression) is proportional to the longitudinal strain $\varepsilon = \frac{\Delta L}{L}$ (elongation or compression per unit length) and Young's modulus E is the proportionality constant.

The above strain is calculated to be

$$\varepsilon = \frac{\sigma}{E} = \frac{50 \cdot 6.8948\ \frac{\text{N}}{\text{mm}^2}}{72900\ \frac{\text{N}}{\text{mm}^2}} \approx 4.73 \cdot 10^{-3} \approx 5\ ‰.$$

With a glass cross-section of $A = 0.0123\ \text{mm}^2$ the tensile force is calculated as

$$F = \sigma \cdot A = 50 \cdot 6.895\ \frac{\text{N}}{\text{mm}^2} \cdot 0.0123\ \text{mm}^2 = 4{,}24\ \text{N}.$$

Table 3.3 Properties of fused silica glass

Property	Unit	Value
Density γ	$\frac{\text{g}}{\text{cm}^3}$	2.20
Young's modulus E	$\frac{\text{N}}{\text{mm}^2}$	72 900
Shear modulus G	$\frac{\text{N}}{\text{mm}^2}$	33 000
Linear thermal expansion coefficient α	K^{-1}	$5.5 \cdot 10^{-7}$

[1]) Kilopound-force per square inch $\left(1\ \text{KPSI} = 6.8948 \frac{\text{N}}{\text{mm}^2}\right)$

▷ *Calculation of Poissons's ratio* μ for a fused silica glass fiber.

Poisson's ratio gives the relationship of cross-sectional strain to longitudinal strain in a body under stress:

$$\mu = \frac{\Delta d}{d} : \frac{\Delta L}{L}$$

By means of Young's modulus E and the shear modulus G it can be calculated as

$$\mu = \frac{E}{2 \cdot G} - 1.$$

Hence for fused silica glass fibers it follows that

$$\mu = \frac{72\,900\ \frac{\text{N}}{\text{mm}^2}}{2 \cdot 33\,000\ \frac{\text{N}}{\text{mm}^2}} - 1 \approx 0.1.$$

In a proof test with a longitudinal strain of approximately 0.5% the calculated diameter decrease is

$$\frac{\Delta d}{d} = \mu \cdot \frac{\Delta L}{L} \approx 0.1 \cdot 5\,‰ = 0.5\,‰.$$

4 Optical Waveguide Profiles

If the refractive index n of an optical waveguide is considered as a function of the radius r, then the term *refractive index profile* is used. It describes the radial change of the refractive index from the axis of the fiber in the core glass outwards toward the cladding glass:

$$n = n\,(r).$$

The propagation of the modes in an optical waveguide depends on the shape of this refractive index profile (Figure 4.1).

The "power law index profiles" are important in practical applications. These are understood to be refractive index profiles, whose curve stated as a power law function of the radius is described as

$$n^2\,(r) = n_1^2\left[1 - 2 \cdot \Delta \cdot \left(\frac{r}{a}\right)^g\right] \qquad \text{for } r < a \text{ in the core}$$

$$\text{and } n^2\,(r) = n_2^2 = \text{constant} \qquad \text{for } r \geq a \text{ in the cladding}$$

where

n_1 refractive index along the axis of the fiber
Δ normalized refractive index difference
r distance from the axis of the fiber in µm
a core radius in µm
g profile exponent
n_2 refractive index of the cladding.

The normalized refractive index difference is related to the numerical aperture NA or the refractive indices n_1 and n_2 as follows:

$$\Delta = \frac{NA^2}{2 \cdot n_1^2} = \frac{n_1^2 - n_2^2}{2 \cdot n_1^2}.$$

The following special cases are worth noting (Figure 4.1):

$g = 1$: triangular profile
$g = 2$: parabolic profile
$g \to \infty$: step profile (limit of g being infinite).

Only in the final instance – with step profile – does the refractive index remain constant, $n(r) = n_1$ in the core glass. For all other profiles, the refractive index $n(r)$ in the core glass rises gradually from the value n_2 of the cladding glass to the value n_1 at the axis of the fiber.

These profiles are therefore called *graded index profiles*. This name has become particularly well established for the parabolic profile (with $g = 2$) because the optical fibers with this profile have technically very good light-guiding qualities.

A further important quantity for the description of an optical fiber is the *V number* or *normalized frequency V*. It is dependent on the core radius a, the numerical aperture NA of the core glass and the wavelength λ or the wave number k of the light. The V number is a dimensionless parameter:

$$V = 2\pi \cdot \frac{a}{\lambda} \cdot NA = k \cdot a \cdot NA,$$

where

a	core radius	NA	numerical aperture
λ	wavelength	k	wave number.

The number N of modes guided in the core glass is dependent on this parameter and is approximately:

$$N \approx \frac{V^2}{2} \cdot \frac{g}{g+2}$$

for an arbitrary power law index profile with the profile exponent g.

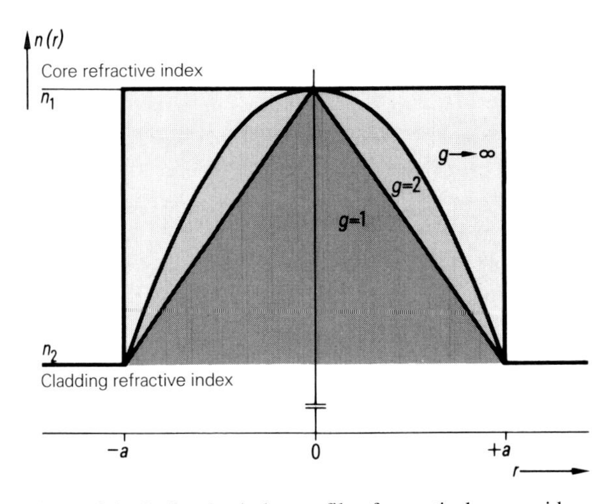

Figure 4.1 Refractive index profile of an optical waveguide

The number of modes in a step index profile ($g \to \infty$) is approximately

$$N \approx \frac{V^2}{2}.$$

For the graded index profile ($g = 2$) the number of modes is

$$N \approx \frac{V^2}{4}.$$

Example

An optical fiber with a graded index profile ($g = 2$), a core diameter $2a = 50$ µm, a numerical aperture $NA = 0{,}2$ and wavelength $\lambda = 1$ µm has a V number

$$V = 2\pi \cdot \frac{\frac{50}{2}\,\mu\text{m}}{1\,\mu\text{m}} \cdot 0.2 = 2\pi \cdot 5 \approx 31.4.$$

The number N of modes guided in the core is then

$$N = \frac{V^2}{4} = \frac{31.4^2}{4} \approx 247.$$

Such an optical waveguide with several modes is called a *multimode fiber.*

If the number of modes is to be decreased, i.e. the V number diminished, either the core diameter $2a$ or the numerical aperture NA must be reduced or the wavelength of the light must be increased. Because the amount of light which can be coupled into an optical fiber is very dependent on the numerical aperture, this value should remain as large as possible. Because of the more difficult handling and jointing, a reduction of the core radius a is limited. On the other hand, it becomes increasingly difficult to build transmitters and receivers for the longer wavelength range; hence the wavelength can not be increased at will.

If the V number in an optical fiber with a step index profile ($g \to \infty$) becomes smaller than the constant $V_{c\infty} = 2.405$, then only a single mode, the *fundamental mode* LP_{01}, can propagate in the core. Such an optical fiber with only one mode is called a *single-mode fiber.*

The value 2.405 is equal to the x value of the Bessel function $J_0(x)$ at its first zero value (Figure 4.2).

The curves of the Bessel functions $J_n(x)$ look like dampened sinusoidal oscillations. They are typical functions for cylindrically symmetrical waveguides such as coaxial cables, hollow tubes or optical fibers.

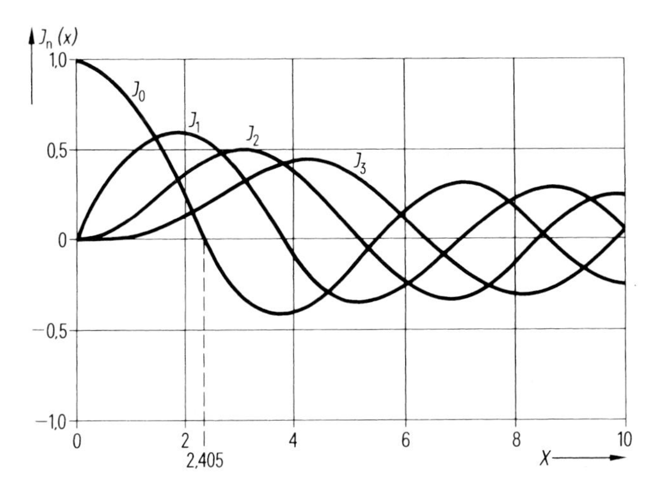

Figure 4.2 Bessel functions

The constant $V_{c\infty}$ is a limit value for an optical waveguide with step index profile ($g \to \infty$). The index c comes from the term cut-off value. For a power law index profile with an arbitrary profile exponent g the limit value V_c can be approximated by

$$V_c \approx V_{c\infty} \cdot \sqrt{\frac{g+2}{g}}.$$

For a graded index fiber ($g = 2$) the limit value V_c is then approximately

$$V_c \approx 2.405 \cdot \sqrt{2} = 3.4\,.$$

Example

An optical fiber with step index profile, a core diameter $2a = 9\ \mu m$ and numerical aperture $NA = 0{,}11$ reaches the V number $V = V_{c\infty}$ at a wavelength

$$\lambda = \pi \cdot \frac{2 \cdot a}{V} \cdot NA = \pi\,\frac{9\ \mu m}{2.405} \cdot 0.11 \approx 1.293\ \mu m = 1293\ nm.$$

The calculated wavelength λ which corresponds to the cut-off value V_c, is called the *cut-off wavelength* λ_c (Chapter 5.5):

$$\lambda_c = \pi \cdot \frac{2 \cdot a}{V_c} \cdot NA.$$

For all wavelengths larger or equal λ_c only one mode can propagate in this specific optical waveguide. This fiber is therefore a single-mode fiber above the wavelength λ_c.

It must be noted here that due to the polarization of light the fundamental mode and all higher modes consist of two modes, whose eigenwaves (modes) oscillate perpendicularly to each other. In spite of the presence of two *polarization modes*, the term single-mode optical waveguide is used. The effects of these polarization modes are not only important for special applications of optical waveguides, e.g. for optical waveguide sensor technology, coherent transmission and for the optical compass, which utilizes "polarization maintaining optical fibers". The effects caused by the split-up into two modes is also important in the communication cable technology for today's transmission systems. On the one hand the generation of so-called composite second order distortions (CSO) due to polarization mode dispersion (PMD) (Chapter 5.10) may cause significant degradation of the quality of analog signals used for CATV applications. On the other hand due to PMD digital signals with bit rates up to 10 Gbit/s used in telecommunication transmission systems may be limited in their link length.

4.1 Step Index Profile

In order that light be guided in the core glass of an optical fiber with *step index profile* due to total internal reflection, the refractive index n_1 of the core glass must be slightly larger relative to the refractive index n_2 of the cladding glass at the interface of the two glasses. If the refractive index n_1 maintains the same value over the entire cross-section of the core, then the refractive index is said to have a step profile. At the interface with the cladding glass the refractive index increases in a step in the core glass and remains unchanged there. Figure 4.3 shows the refractive index profile of an optical fiber with step index profile. It also shows the propagation of a light ray with the corresponding angles.

Such an optical waveguide is termed a fiber with step index profile or *step index fiber*. The following example is chosen to better illustrate the propagation of light in such an optical fiber (Figure 4.3).

Typical Dimensions for a Multimode Fiber with Step Index Profile

Core diameter	$2a$	100 µm
Core refractive index	n_1	1.48

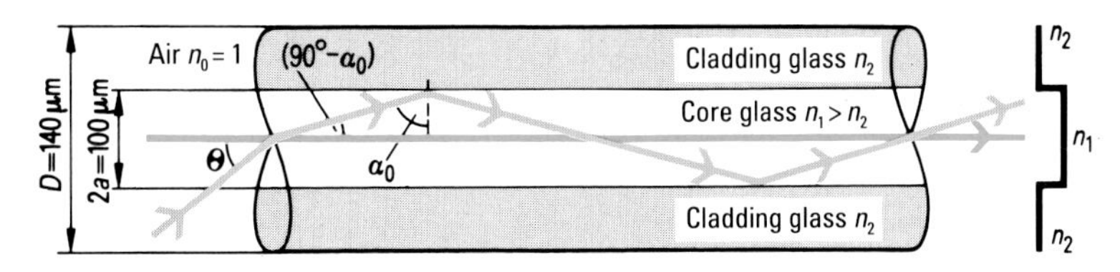

Figure 4.3 Fiber with step index profile

Cladding diameter	D	140 µm
Cladding refractive index	n_2	1.46.

In this case, the critical angle α_0 of total internal reflection, i.e. the smallest angle between the incident light ray and the normal under which a light ray is guided in the core glass and not refracted into the cladding glass, is

$$\sin \alpha_0 = \frac{n_2}{n_1} = \frac{1.46}{1.48} \approx 0.9865;$$

$$\alpha_0 \approx 80.6°.$$

All light rays which form an angle with the fiber axis smaller than or equal to $(90° - \alpha_0) = 9.4°$ are guided in the core glass.

When light is launched in from outside (air $n_0 = 1$) into the core glass, the law of refraction must be taken into consideration, because light can only enter the fiber within a specific acceptance angle Θ. In this case the angle is

$$\sin \Theta = \sqrt{n_1^2 - n_2^2} = \sqrt{1.48^2 - 1.46^2} \approx 0.242;$$

$$\Theta \approx 14.0°.$$

Because the sine of the acceptance angle is defined as the numerical aperture, the result for the NA is

$$NA = \sin \Theta \approx 0.242.$$

The normalized refractive index difference Δ for this fiber with step index profile is

$$\Delta = \frac{NA^2}{2 \cdot n_1^2} \approx \frac{0.242^2}{2 \cdot 1.48^2} \approx 0.0134 = 1.34\%.$$

The V parameter calculated for a step index fiber with core diameter $2a$ = 100 μm at a wavelength λ = 850 nm is

$$V = \pi \cdot \frac{2a}{\lambda} \cdot NA = \pi \cdot \frac{100\ \mu\text{m}}{0.85\ \mu\text{m}} \cdot 0.242 \approx 89.4.$$

The number N of modes for such an optical fiber is then approximately

$$N \approx \frac{V^2}{2} = \frac{89.4^2}{2} \approx 4000.$$

Such an optical fiber is a multimode fiber. A light pulse propagating in it is composed of many partial pulses which are guided in the individual modes of the fiber. Each of these modes is excited at the beginning of the fiber with a different launch angle and is guided through the core of the fiber in a correspondingly different ray path. Each mode travels a different distance in the process and therefore arrives at different times at the end of the fiber. The longest transit time is related to the shortest comparable to the ratio of the refractive index of the core and cladding glass and therefore amounts roughly to the dimension of the refractive index difference Δ of more than 1 per cent.

Example

Light travels through a fiber with step index profile 1 km in length in approximately 5 μs. The delay time (transit time difference) is then about

$$\Delta t \approx 5\ \mu\text{s} \cdot 0.01 = 50\ \text{ns}.$$

The delay time distortion of the individual modes is called *modal dispersion*. It causes a short light pulse to broaden in time as it passes through a step index fiber. This is disadvantageous for optical transmission, because it reduces the transmission speed (bit rate) and the transmission bandwidth. This effect is moderated because the individual modes act on each other and exchange energy along the route. This *mode mixing or mode coupling* occurs with special intensity at disturbed points in the core glass, e.g. at splices and bends of the fiber.

During the course of the modes along the axis of the fiber, low-order modes with a small angle to the axis of the fiber are transformed by energy exchange to modes of a higher order with a steeper angle to the axis of the fiber, and vice versa. Thereby the difference in speeds of the modes is evened out, with the result that the time broadening Δt in the launched light pulse does not increase linearly with the fiber length L, that is $\Delta t \sim L$, but rather ideally only with the square root of the length, that is $\Delta t \sim \sqrt{L}$.

This modal dispersion can be eliminated altogether when the step index fiber is dimensioned such that only one mode is guided, namely the fundamental mode LP_{01}.

However, the fundamental mode also broadens in time as it passes through such a fiber. This effect is called *chromatic dispersion* (Chapter 5.4). Because it is a property of the material, it generally occurs in every optical fiber. In comparison to the modal dispersion the chromatic dispersion is relatively minor or zero in the wavelength range from 1200 to 1600 nm.

The term *mode field diameter* $2w_0$ has been introduced to quantify the size (radial field amplitude) of the fundamental mode (Chapter 5.6). In order to manufacture a low-attenuation step index fiber which guides only the fundamental mode in the wavelength range above 1200 nm, the mode field diameter $2w_0$ must be reduced to about 9 µm. Such a step index fiber is termed a *single-mode fiber* or *monomode fiber*.

Typical Dimensions for a Single-Mode Fiber

Mode field diameter	$2w_0$	9 µm
Cladding diameter	D	125 µm
Core refractive index	n_1	1.46
Refractive index difference	Δ	0.003 = 0.3%.

Figure 4.4 shows the path of light and the refractive index profile of a single-mode fiber.

The numerical aperture NA of a typical single-mode fiber is

$$NA = n_1 \cdot \sqrt{2\Delta} \approx 1.46 \cdot \sqrt{2 \cdot 0.003} \approx 0.113$$

with an acceptance angle of Θ of

$$\sin \Theta = NA \approx 0.113$$
$$\Theta \approx 6.5°.$$

It is useful to note here that not only the core diameter but also the numerical aperture and therefore the acceptance angle too are much smaller in comparison to the multimode step index fiber, which makes it comparatively more difficult to launch light into the single-mode fiber.

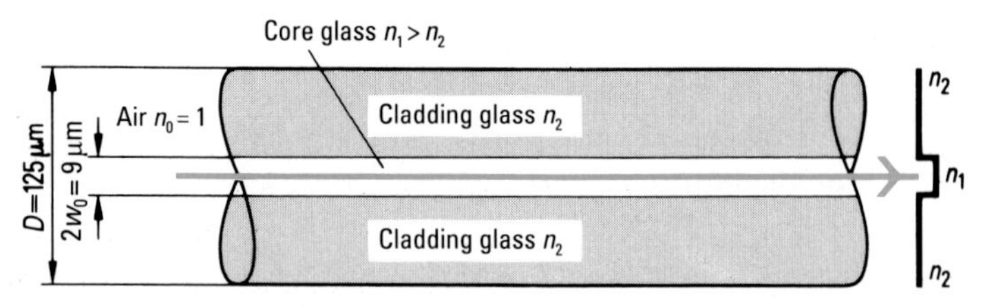

Figure 4.4 Single-mode fiber

The cut-off wavelength λ_c, above which only the fundamental mode is guided in the fiber and at which the V parameter is $V_c = 2.405$, is calculated for a typical single-mode fiber as

$$\lambda_c = \pi \cdot \frac{2a}{V_c} \cdot NA = \pi \cdot \frac{8{,}5\ \mu m}{2{,}405} \cdot 0.113 \approx 1.255\ \mu m = 1255\ nm.$$

At this wavelength, λ_c, the next higher mode LP_{11} (Figures 2.9 and 4.6) can no longer propagate in the fiber. Only the fundamental mode LP_{01} is still present at higher wavelengths. However, its mode field also broadens into the cladding glass (Figure 4.5).

Example

A single-mode fiber with step index profile, a core diameter $2a = 8.5\ \mu m$ and a cut-off wavelength $\lambda_c = 1255$ mn ($V_c = 2.405$) has a mode field diameter $2w_0$ at the wavelengths $\lambda = 1300$ and 1550 nm:

$$2 \cdot w_0 \approx \frac{2.6 \cdot \lambda}{V_c \cdot \lambda_c} \cdot 2a \text{ (Chapter 5.6)}$$

$$\lambda = 1300\ \text{nm: } 2 \cdot w_0 \approx \frac{2.6 \cdot 1300\ nm}{2.405 \cdot 1255\ nm} \cdot 8.5\ \mu m \approx 9.5\ \mu m;$$

$$\lambda = 1550\ \text{nm: } 2 \cdot w_0 \approx \frac{2.6 \cdot 1550\ nm}{2.405 \cdot 1255\ nm} \cdot 8.5\ \mu m \approx 11.3\ \mu m.$$

When single-mode fibers are bent or spliced, the size of the mode field diameter is an important factor for the attenuation characteristics. A larger mode field diameter causes poorer light guiding in bends but lower losses with splice and connector joints.

4.2 Graded Index Profile

In a step index fiber with numerous modes, these modes propagate along paths of varying length and therefore arrive at different times at the end of the fiber. This undesirable modal dispersion can be greatly reduced, when the refractive index of the core glass diminishes parabolically from a maximum value n_1 at the axis of the fiber to a refractive index n_2 at the interface with the cladding. Such a graded index profile or power law profile with the profile exponent $g = 2$ is defined by

$$n^2(r) = n_1^2 - NA^2 \cdot \left(\frac{r}{a}\right)^2 \qquad \text{for } r < a \text{ in the core}$$

and $$n^2(r) = n_2^2 \qquad \text{for } r \geq a \text{ in the cladding.}$$

An optical waveguide with this graded index profile is also called a *graded index fiber.*

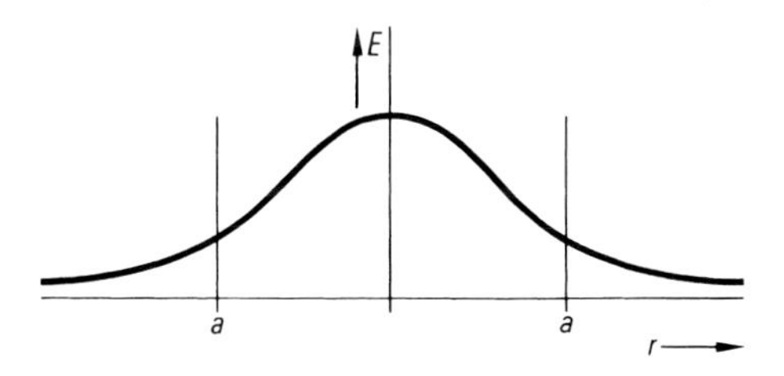

Figure 4.5 Radial field distribution of the fundamental mode LP_{01}

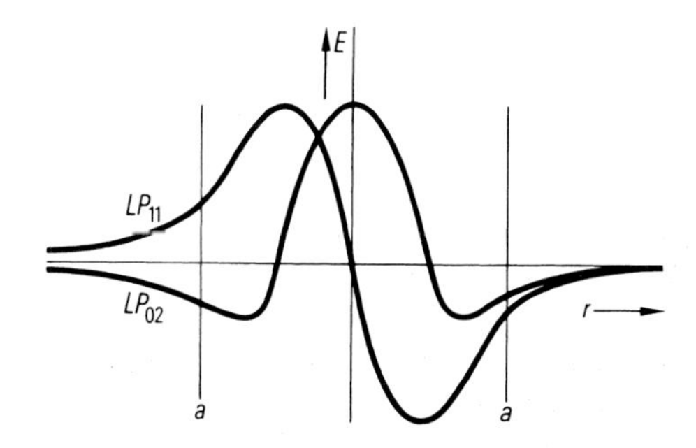

Figure 4.6 Radial field distribution of the modes LP_{11} and LP_{02}

Typical Dimensions of a Fiber with Graded Index Profile

Core diameter	$2a$	50 µm
Cladding diameter	D	125 µm
Maximum core refractive index	n_1	1.46
Refractive index difference	Δ	0.010.

Figure 4.7 shows the course of light waves of different order and the refractive index profile of a fiber with graded index profile.

The light rays pass through the optical fiber on wave- or screw-like helical paths. In contrast to the step index profile, they no longer propagate in zig-zags. Due to the continuous change in the refractive index $n(r)$ in the core glass, the rays are refracted continuously and hence their direction of propagation is changed, thereby propagating in wave paths. The rays oscillating around the fiber axis still travel a longer path than the light ray along the fiber axis; however, due to the lower refractive index outside of the fiber axis these rays travel correspondingly faster, which evens out the time delay for the longer path. The result is that the delay time difference of the various rays disappears almost completely. When the parabolic shape of the refractive index profile is manufactured very precisely in a graded index fiber the delay time differences will only be slightly more than 0.1 ns for a transit time of light of 5 µs over 1 km.

This minimal delay time difference in graded index fibers is caused by "profile dispersion" in addition to the material dispersion. It is due to the different variations in the refractive indices of the core and cladding when the wavelength λ changes, and therefore the refractive index difference Δ and the profile exponent g are wavelength dependent. The optimal profile exponent of a parabolic graded index profile can be calculated on a theoretical basis from

$$g = 2 - 2P - \Delta \cdot (2 - P),$$

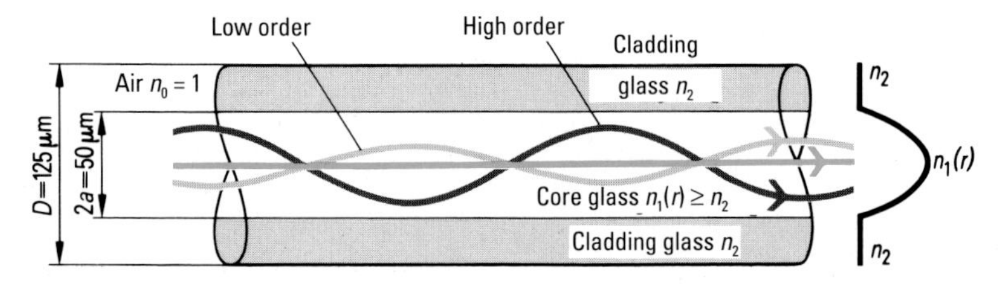

Figure 4.7 Fiber with graded index profile

whereby both the parameter $P \ll 1$ and the refractive index difference Δ are dependent on the wavelength λ and therefore also the profile exponent g.

It is only within a limited wavelength range that the refractive index profile of an optical fiber which has a graded index profile with $g \approx 2$ makes it possible for all guided modes to have almost the same delay time.

Because the refractive index $n(r)$ in an optical fiber with graded index profile is dependent on the distance r from the fiber axis, the acceptance angle Θ, which is important for launching light into the core, is a function of r:

$$\sin \Theta(r) = \sqrt{n_1^2(r) - n_2^2} = NA \cdot \sqrt{1 - \left(\frac{r}{a}\right)^2} \leq NA.$$

The acceptance angle is largest at the fiber axis ($r = 0$); to be precise it is equal to the numerical aperture NA. At the interface with the cladding ($r = a$) the angle is equal to zero.

For a typical fiber with graded index profile the numerical aperture is

$$NA = n_1 \cdot \sqrt{2\Delta} \approx 1.46 \cdot \sqrt{2 \cdot 0.01} \approx 0.206$$

and the maximum acceptance angle Θ_{max} at the axis of the fibers is

$$\sin \Theta_{max} = NA \approx 0.206;$$
$$\Theta_{max} \approx 11.9°.$$

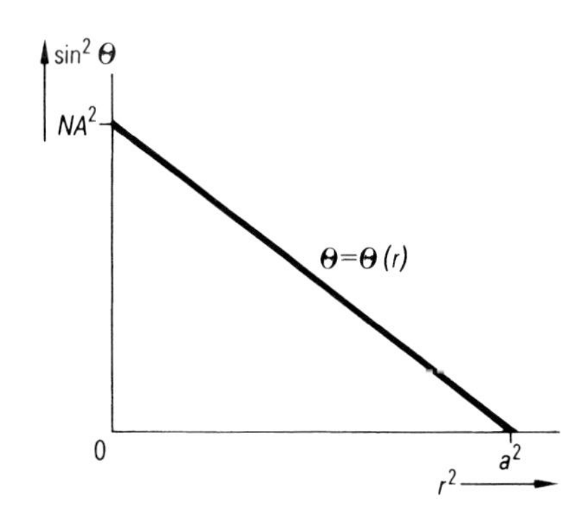

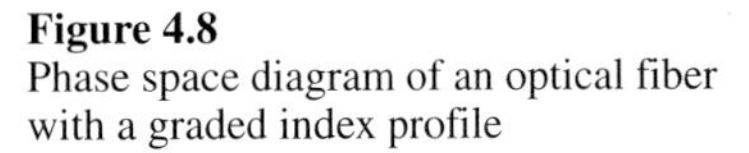

Figure 4.8
Phase space diagram of an optical fiber with a graded index profile

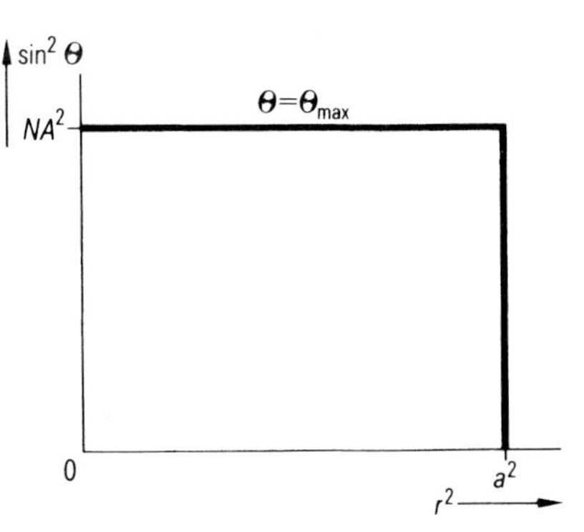

Figure 4.9
Phase space diagram of an optical fiber with a step index profile

For plotting the acceptance angle Θ as a function of the distance r from the axis of the optical fiber, the coordinates $\sin^2 \Theta$ and r^2 have proved to be most useful. A system with these coordinates is called a *phase space diagram*. Figures 4.8 and 4.9 show diagrams for an optical fiber with graded index profile and for purpose of comparison for an optical fiber with step index profile.

The area limited by the curve of the maximum acceptance angle Θ_{max} is proportional to the light power which can be launched into the core. As can be seen, with the same numerical aperture NA and the same core radius a this power is twice as high in an optical fiber with step index profile as it is in one with graded index profile. The number N of modes guided in the core is also proportional to this area. The individual modes can be localized inside this area. For example the lower order modes $LP_{\nu\mu}$, whose mode numbers are approximately $\nu = 0, 1, 2$ and $\mu = 1, 2$ and which propagate almost parallel to the optical fiber axis, are close to the origin. The higher order modes $LP_{\nu\mu}$ (with $\nu, \mu \gg 1$) are farther from it (Chapter 2.7). Modes which lie outside this area are not guided, i.e. they have very high attenuation. Modes which lie only slightly above the limiting curve still have some ability to propagate but show heightened attenuation. Such modes are called *leaky modes*; they are partially guided and partially radiated (Figure 4.10).

Other optical fiber properties can be described by means of the phase space diagram, as, for example, launch conditions (Chapter 5.1); it is also used to calculate the light power which can be launched into an optical fiber from a light source.

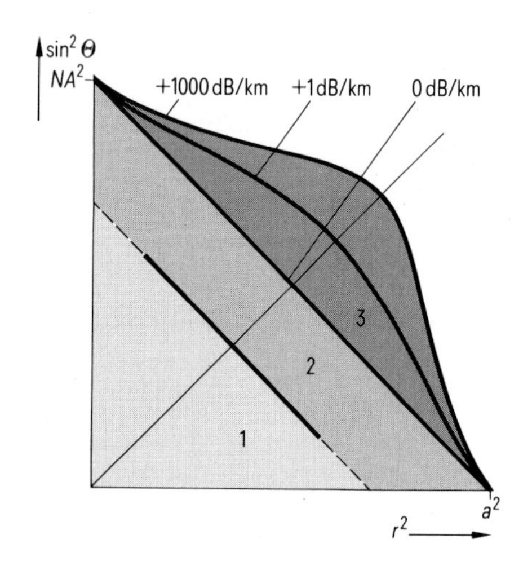

Figure 4.10
Modes in the phase space diagram

1 Lower order modes
2 Higher order modes
3 Range of leaky modes

4.3 Multistep Index Profile

The dispersion in an single-mode optical fiber is composed of two types of dispersion. On the one hand, there is the material dispersion caused by the wavelength dependence of the refractive index $n = n\ (\lambda)$ (Figure 2.6) and therefore of the light speed $c = c\ (\lambda)$. On the other hand, there is the waveguide dispersion which results from the wavelength dependence of the light distribution of the fundamental mode LP_{01} over core and cladding glass (Figure 4.5) and therefore of the refractive index difference $\Delta = \Delta\ (\lambda)$. Both types of dispersion taken together are called chromatic dispersion.

In the wavelength range higher than 1300 nm the two dispersions in fused silica glass have opposite signs. The material dispersion can only slightly be changed through the use of other glass dopants. By contrast the waveguide dispersion can be greatly influenced by the use of another profile structure of the refractive index.

The refractive index profile of a typical single-mode fiber is a step index profile with the refractive index difference Δ. For this simple profile structure the sum of the material and waveguide dispersions is zero near the wavelength $\lambda = 1300$ nm.

If it is desirable to move this zero dispersion wavelength toward different wavelengths, then the waveguide dispersion and with it the profile structure

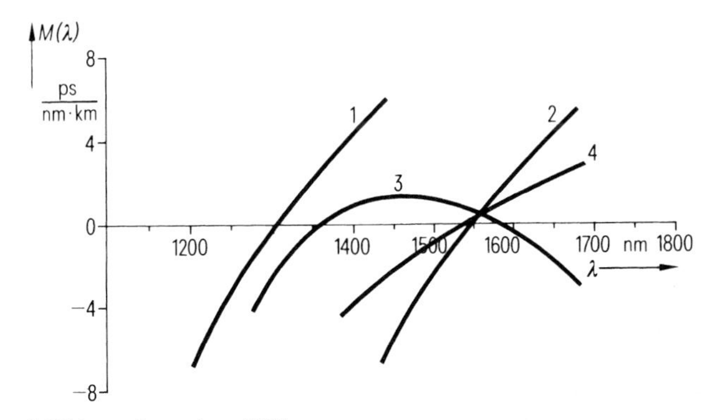

1 Without dispersion shifting
2 With dispersion shifting
3 With dispersion flattening
4 With nonzero dispersion shifting

Figure 4.11 Chromatic dispersion as a function of the wavelength

of the optical fiber must be changed. This leads to the *multistep* or *segmented index profiles*. By using these profiles, it is possible to produce optical fibers whose zero dispersion wavelength is shifted towards 1550 nm (so-called *dispersion shifted fibers*) or whose dispersion values are very low for the entire wavelength range from 1300 to 1550 nm (so-called *dispersion flattened or dispersion compensated fibers*) or whose dispersion values are low but nonzero over the wavelength range from 1530 nm to 1625 nm (so-called *nonzero dispersion shifted fibers*).

Figure 4.11 shows the chromatic dispersion $M(\lambda)$ as a function of the wavelength λ for a single-mode fiber without dispersion shifting (1), with dispersion shifting (2), with dispersion flattening (3) and with nonzero dispersion shifting (4).

These single-mode fibers can be produced with different profile design. The following provides a selection of these profiles.

Type 1 Without dispersion shifting

Standard step index profile (simple step index or matched cladding) (Figure 4.12a)

Step index profile with reduced refractive index in the cladding (depressed cladding) (Figure 4.12b)

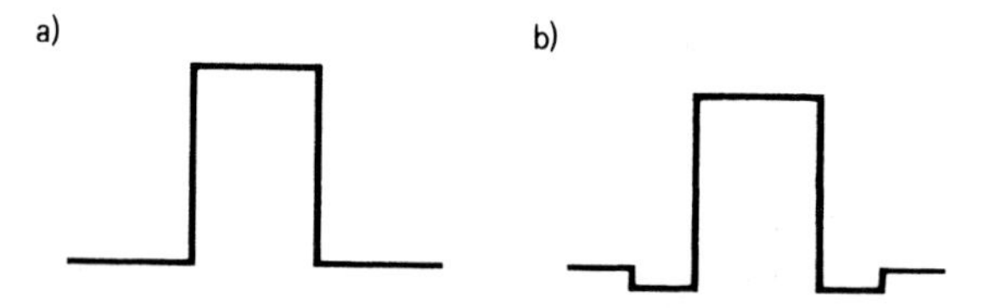

Figure 4.12 Profile design of optical fibers without dispersion shifting

Type 2 With dispersion shifting, Type 4 With nonzero dispersion shifting

Segmented profile with triangular core (segmented core) (Figure 4.13a)

Triangular profile (Figure 4.13b)

Segment profile with a double step refractive index in the cladding (double clad) (Figure 4.13c)

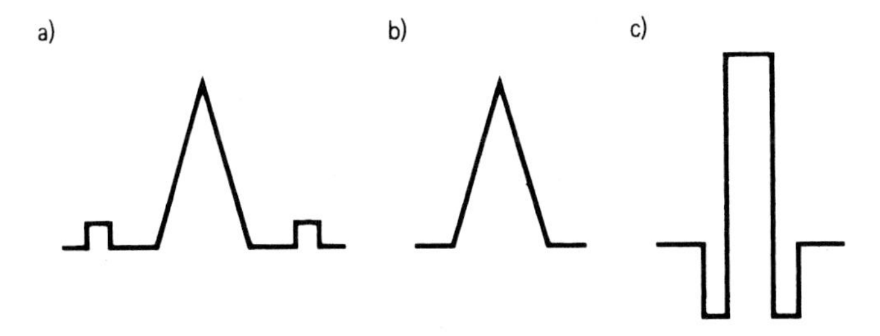

Figure 4.13
Profile designs of optical fibers with dispersion shifting and nonzero dispersion shifting

Type 3 With dispersion flattening

Segmented profile with a fourfold step in refractive index of the cladding (quadruple clad) (Figure 4.14a)

W profile (double clad) (Figure 4.14b)

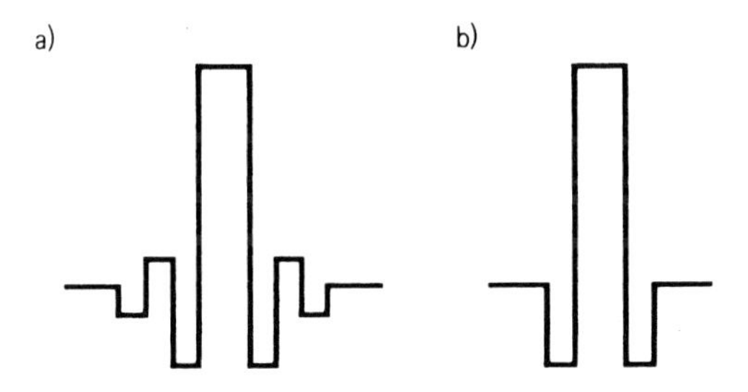

Figure 4.14
Profile design of optical fibers with dispersion flattening

5 Optical Fiber Parameters and Measurement Methods

The quality of a fiber optic cable is tested using established measurement methods. Standards need to be set for the optical fiber parameters and the corresponding measurement methods. At the European level, the Subcommittee SC 86A of the Comité Européen de Normalisation Electrotechnique (CENELEC) is responsible for these standards, at the national level the Subcommittee UK 412.6 of the German Electrotechnical Commission (DKE, Deutsche Elektrotechnische Kommission) and at the international level the Technical Committee TC 86 of the International Electrotechnical Commission (IEC).

The technical report IEC 61931, Fiber Optic Terminology, provides the terminology for the various fiber optic elements, devices and systems.

5.1 Launching Conditions

The procedure for launching light into an optical fiber is very important in the further distribution of the light power in the fiber, because in multimode fibers the power of the launched light pulses is distributed over the individual modes. In the case of single-mode fibers light is partly launched into the fundamental mode and partly radiated. Depending on the launching conditions, when light is launched into single-mode and multimode fibers, they will guide the light power differently.

5.1.1 Multimode Fibers

During *full flood launch* the entire core of the optical fiber is flooded with light, which excites all guided modes of higher and lower order and also the leaky modes[1)] (Figure 4.10). Because these modes have varying degrees of attenuation along the optical fiber and in addition cause *mode mixing* due to en-

[1)] Radiation waves, which are caused by continuous partial radiation of energy out of a waveguide.

ergy transfer a different distribution of the light power and the delay time of the modes dependent on the length of the optical fiber will be measured. The conditions at the end of the multimode fiber therefore depend on the launching conditions at the beginning – quite independent of a full flood or non-full flood launch – and on the mode mixing along the path.

For multimode fibers it is particularly important that an unambiguous method of light launching be defined so that precise and reproducible measurement methods for the most important transmission parameters can be developed. For this purpose a basic premise is that dependent on the degree of mode mixing a *steady state* or *equilibrium mode distribution* (EMD) is reached after a certain length in a multimode fiber, beyond which the energy distribution remains constant over the modes.

It therefore becomes particularly desirable to measure transmission parameters in this steady state along the optical fiber. There are different technical methods for reaching this state. One possibility is to use a *dummy fiber* of sufficient length which has already reached the equilibrium mode distribution to launch light into the optical fiber under test. Because the equilibrium mode distribution is only achieved after a very long length, these dummy fibers would have to be impractically long. This can be avoided by using a shorter length of optical fiber in which statistically irregular mechanical perturbations induce strong mode mixing, thereby approximating equilibrium mode distribution in this piece of optical fiber. Such setups are called *mode scramblers*. For this purpose the short length of optical fiber can be pressed against a rough surface (e.g. sandpaper or files) or bent around little balls. When an optical fiber with step index profile is spliced to one with graded index profile to which an optical fiber with step index profile is spliced (*SGS arrangement*) each segment being 1 to 2 m in length, intense mode coupling also results.

If higher order modes are to be suppressed at launch a *mode filter* is used. This involves, for example, wrapping the optical fiber around a mandrel approximately 1 cm in diameter which strips off the higher order modes. Generally speaking, a mode scrambler can be used to excite all modes and a mode filter to limit the excitation to certain (in particular lower order) modes.

In addition to these mechanical methods optical aids are also often used to achieve an equilibrium mode distribution. By exciting only low-order modes, problems with leaky modes and cladding modes can be avoided. To this end a suitable combination of lenses and apertures can be used to create an input light ray which fills 70 per cent of the core diameter and 70 per cent of the numerical aperture. This launch lends itself well to graphical documentation in a phase space diagram. As an example Figure 5.1 graphs for a graded index

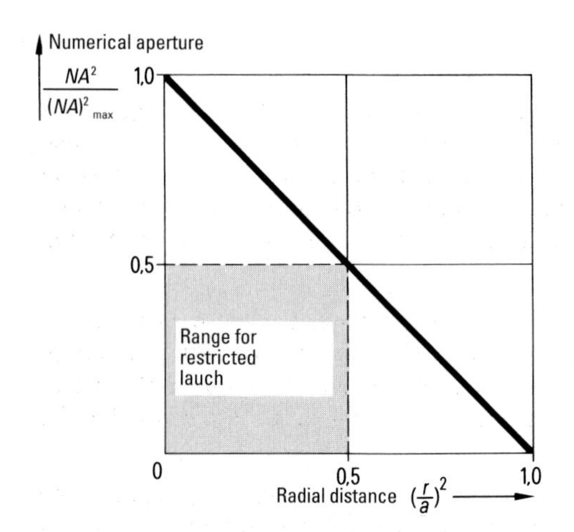

Figure 5.1
Phase space diagram of a restricted launch into a graded index fiber

r Radial distance

a Fiber core radius

fiber the range in which the modes are excited during such a restricted or limited phase space (LPS) launch.

In order to determine whether a launch creates an equilibrium mode distribution or comes to this state, the near and far field distribution of the particular configuration must be measured. Experience indicates that in the case of an optical fiber with graded index profile – with a core diameter of 50 µm – equilibrium mode distribution is approximated when the bell-shaped radial near field light distribution (Chapter 5.7) has a full half width of approximately 26 µm after 2 m and the respective far field light distribution, which is also bell-shaped, as a function of the angle reaches half widths corresponding to a numerical aperture of 0.11.

5.1.2 Single-Mode Fibers

During full flood launch into a single-mode fiber cladding modes are generated that, however, are stripped off within a few centimeters when a coating is used that has a refractive index larger than that of the cladding glass. This coating acts as a *mode stripper.*

5.2 Attenuation

Light propagating in an optical fiber is attenuated, i.e. looses energy. This loss should be kept to a minimum, so that great distances can be bridged without insertion of repeaters. The attenuation of an optical fiber is an important parameter in dimensioning an optical communication cable plant. It is caused mainly by the physical processes, absorption and scattering (Chapter 3.1.1).

The magnitude of the light losses depends among other things upon the wavelength of the launched light. Therefore it is generally useful to measure the attenuation of an optical fiber spectrally, i.e. as a function of the wavelength, in order to find wavelength ranges with low attenuation suitable for optical communication.

Whereas the absorption process only takes place at certain wavelengths – the so-called absorption bands (e.g. at 1390 nm the OH absorption) – light loss due to scattering occurs at all wavelengths. Because scattering in optical fibers is the result of density fluctuations (inhomogeneities) with dimensions mostly smaller than the wavelength of the light, the *Rayleigh scattering law* can be used to describe this process. It states that as the wavelength λ increases the scattering loss α decreases with the fourth power of λ (Figure 5.2).

$$\alpha \sim \frac{1}{\lambda^4}.$$

By comparing the scattering losses, e.g. at the wavelengths 850, 1300 and 1550 nm, which are important for optical communication, it can be seen that at 1300 nm the scattering loss reaches only 18 per cent of its value at 850 nm

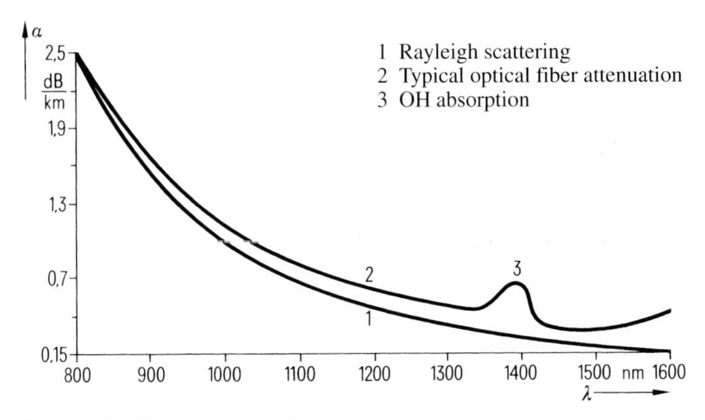

Figure 5.2 Rayleigh attenuation curve

and at 1550 nm only 9 per cent of it. Operation of optical cables at these wavelengths is therefore advantageous.

When the light propagation in a steady state is considered, it can be seen that guided light power P decreases exponentially with the length L of the optical fiber:

$$P(L) = P(0) \cdot 10 - \alpha \frac{L}{10} .$$

$P(0)$ is the light power which is launched into the optical fiber at the beginning, $P(L)$ is the light power remaining after the length L and α is the attenuation coefficient which is a measure of the attenuation per unit length. An optical fiber with the length L and the attenuation coefficient α has the attenuation

$$\alpha \cdot L = 10 \cdot \log \frac{P(0)}{P(L)},$$

where

α attenuation coefficient in $\frac{\text{dB}}{\text{km}}$

L length of the fiber in km.

Example

An attenuation of 10 dB means that the light power $P(L)$ in an optical fiber after the length L (in km) is only 10 per cent of the light power $P(0)$ at the beginning of the fiber; at 3 dB it is still 50 per cent and at 1 dB it is approximately 80 per cent.

Modern single-mode fibers have an attenuation of 0.2 dB per kilometer at a wavelength of 1550 nm, i.e. only 4.5 per cent of optical power is lost over one kilometer. The spectral curve of the attenuation coefficient of a typical single-mode and multimode fiber is shown in Figure 5.3.

In order to determine the attenuation coefficient of an optical fiber the light power must be measured at two points of the fiber. Thereby a steady state in the fiber must be given; i.e. the launching of light must be carried out carefully in such a way that between the measuring points no cladding light remains in a single-mode fiber and a state of equilibrium mode distribution is achieved in a multimode fiber. Therefore the limited phase space launch (70 per cent) is mainly used for the measurement of the attenuation coefficient.

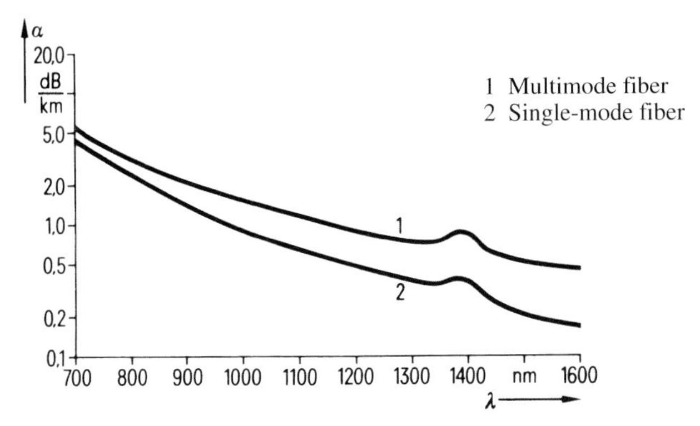

Figure 5.3
The spectral curve of the attenuation coefficient of a single-mode and multimode fiber

5.2.1 Measurement Methods

Through Power Technique

There are two methods of carrying out the through power technique (Figure 5.4), the cut-back method and the insertion loss technique.

The *cut-back method* requires measurement of the optical power at two points L_1 and L_2 of the optical fiber. Usually L_2 is at the end of the fiber and L_1 very near the beginning. In the measuring process the light power P is first measured at the end at L_2 (in km) and then again at L_1 (in km), which involves cutting the fiber back to the point L_1 without changing the launch conditions between the light source (transmitter) and the fiber. The attenuation coefficient $\alpha \left(\text{in} \dfrac{\text{dB}}{\text{km}} \right)$ of the fiber is then calculated by

$$\alpha = \frac{10}{L_2 - L_1} \cdot \log \frac{P(L_1)}{P(L_2)} .$$

This method is not without problems, for it requires that a short length of the optical fiber be cut off, which is not practical for connectorized optical cables. The *insertion loss technique* is useful in this case. It requires that the light power at the far end of the fiber under test be measured and then compared with the light power at the end of a short piece of optical fiber. This piece of fiber is used as reference and should be similar in structure and properties of the fiber under test. As the test is being carried out, it is important to take care that the launch conditions for the reference length are as similar as possible to those for the length being measured. Due to these limitations, accuracy and reproducibility of the insertion loss technique are less favorable than with the cut-back method.

With the through power technique (Figure 5.4), light is launched into the beginning of an optical fiber (at the transmitter) and then passes through the fiber to be measured at the receiver end.

It can be a disadvantage because this is a total measurement over the entire fiber length and gives no details about localized attenuation changes along the optical fiber. Furthermore, both ends of the optical fiber must be accessible.

Bending Coupler Technique

The bending coupler technique is another method for measuring attenuation. It requires only a few centimeters of optical fiber for coupling light into and out of the fiber, this means that it is not necessary to have access to the ends of the optical fiber. This technique therefore permits quick and simple measurement of loss due to splices, couplers and optical splitters or wavelength multiplexers.

Backscattering Technique

In the backscattering technique light is launched into as well as received at one end of the fiber (Figure 5.5). In addition it also provides details of the attenuation along the length of the optical fiber.

Rayleigh scattering provides the basis for this technique. At the same time as the larger part of light power propagates forwards, a small percentage is reflected backwards towards the transmitter. This reflected power is also attenuated as it propagates back through the fiber. The remaining power is coupled out by a beam splitter, e.g. a partly transparent mirror, and measured. From

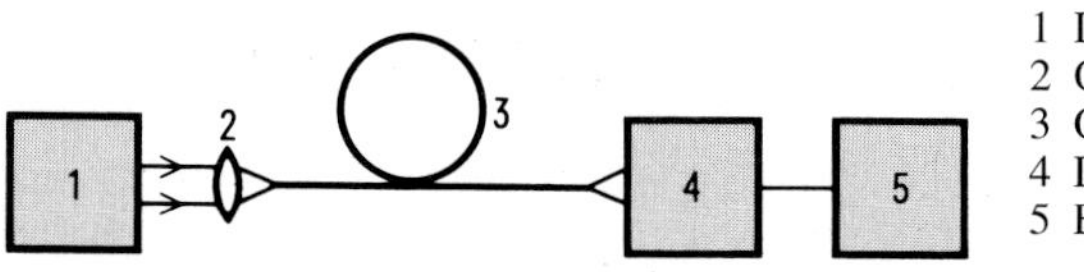

1 Light source
2 Optics
3 Optical fiber
4 Detector
5 Evaluation

Figure 5.4 Through power technique

this backscattered light power (Figure 5.6) and the transit time in the optical fiber it is possible to get a curve from which the attenuation coefficient over the entire length of the fiber can be derived. The course in time of the backscattered signal can easily be observed on an oscilloscope.

If the attenuation coefficient and the backscattering factor remain constant over the fiber length, then the curve falls exponentially from the beginning of the optical fiber. Due to the jump in refractive index at the beginning and at the end of the fiber a relatively large portion of the light power is backscattered, there, which causes the peak at the beginning and end of the curve. From the time difference Δt between the two peaks, the speed of light in vacuum c_0 and the group refractive index $n_g \approx 1.5$ in the core glass, the length L of the fiber can be calculated:

$$L = \frac{\Delta t}{2} \cdot \frac{c_0}{n_g}.$$

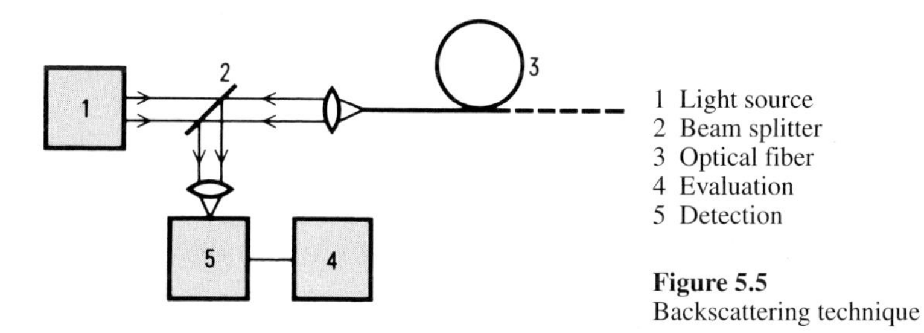

1 Light source
2 Beam splitter
3 Optical fiber
4 Evaluation
5 Detection

Figure 5.5
Backscattering technique

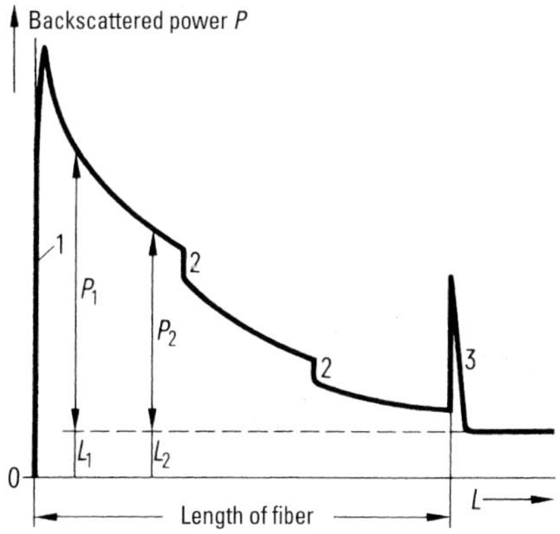

Backscattered power P

P_1 P_2 L_1 L_2 0 Length of fiber L

1 Backscattering at beginning of fiber
2 Backscattering at connecting splice
3 Backscattering at end of fiber

Figure 5.6
Backscattering measurement curve

L length of the fiber in km
Δt time difference between the peaks of the initial and end pulses in s

c_0 speed of light in vacuum $300\,000\ \frac{\text{km}}{\text{s}}$

n_g effective group refractive index in the core glass.

The attenuation coefficient α for any partial segment of the optical fiber between L_1 and L_2 is calculated by

$$\alpha = \frac{5}{L_2 - L_1} \cdot \log \frac{P(L_1)}{P(L_2)}.$$

Due to the fact that the light travels forward and backward the factor 5 is used here in contrast to that in the equation for the through power technique. This equation applies assuming that the backscattering factor, the numerical aperture and the core diameter remain unchanged over the length of the optical fiber. If this is not assured, two measurements should be made, one from each end, and the results averaged. Because the backscattered power is relatively low, greater demands are placed on the sensitivity of the receiver.

In order to improve the received signal, the individually measured values are averaged many times. Test units operating on the backscattering principle are called *optical time domain reflectometers* (OTDR). In addition to the measurement of the attenuation coefficient, it can also be used to locate perturbations (fractures) in the fiber and to check light losses in jointed fibers (jumps in attenuation due to connector or splice joints).

The through power measurement technique is specified in European standard EN 188 000 (national German standard VDE 0888 Part 101) and internationally in IEC 60793-1-4 methods C1A / C1B.

The backscattering measurement technique is specified in European standard EN 188 000 (national German standard VDE 0888 Part 101) and internationally in IEC 60793-1-4 method C1C.

5.3 Bandwidth

For the description of the transmission characteristics of an optical fiber, the second important parameter used in addition to the attenuation α is the bandwidth B or the bandwidth length product b_1 (Chapter 10.1.2). On the one hand, the attenuation gives the loss of light power along an optical fiber; on the other hand, the bandwidth is used as a measure for the dispersion properties of an optical fiber.

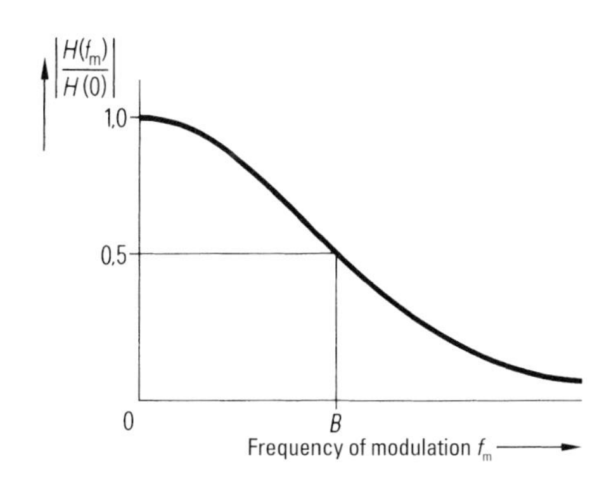

Figure 5.7
Transfer function of an optical fiber

Due to the dispersion (Chapter 5.4) light pulses broaden in time when they propagate through an optical fiber. In terms of frequency, this means that an optical fiber acts as a low-pass filter.

In other words, in a fiber the amplitude of a light wave decreases as the *frequency of modulation* f_m increases, until it finally disappears. Therefore the optical fiber permits low-frequency signals to pass and attenuates them as the frequency increases. By measuring for each frequency of modulation the amplitudes of the light power at the beginning $P_1(f_m)$ and end $P_2(f_m)$ of a fiber and by calculating the ratio of these amplitudes, the value of the *transfer function* $H(f_m)$ can be derived:

$$H(f_m) = \frac{P_2(f_m)}{P_1(f_m)}.$$

$H(f_m)$ is a function of the frequency of modulation f_m. Usually the magnitude of the transfer function is normalized by dividing it by $H(0)$. $H(0)$ is the magnitude of the transfer function at the frequency $f_m = 0$ Hz, i.e. without modulation. Figure 5.7 shows a typical curve for it.

The form of this bell-shaped curve roughly approximates a Gaussian low-pass filter. The frequency of modulation at which the normalized magnitude of the transfer function equals 0.5 is called the bandwidth B of the optical fiber:

$$\frac{H(f_m = B)}{H(0)} = 0.5.$$

Therefore the bandwidth is the frequency of modulation at which the amplitude (light power) relative to the value at the frequency zero has fallen by 50 per cent or 3 dB optically.

5.3.1 Measurement Methods

For multimode fibers the measurement of the bandwidth is dependent on the launch conditions of the modes, i.e. on the conditions under which light is launched into the optical fiber. If, for example, light with a rather small aperture is spotlighted onto a small area of the core, then only the modes there are excited and their specific individual delay times can be studied. This technique is called *differential mode delay* (DMD) measurement. It allows the investigation of the refractive index profile and whether it approaches the ideal form without allowing any modal dispersion. Perturbations in the profile are indicated by time delays in the corresponding modes, which in turn results in pre- or post-pulses in time.

Due to mode mixing (Chapter 5.1) all propagative modes achieve an equilibrium mode distribution – dependent on the properties of the optical fiber and the external conditions. Although attenuation measurements result in values which are proportional to the length of the optical fiber, this is only true under certain conditions for the bandwidth measurement. The dependence of the bandwidth B on the length L of the fiber can be approximated by the γ factor (also called the length exponent), which itself, however, depends on the length. The following holds:

$$\frac{B}{B_1} = \left(\frac{L}{L_1}\right)^{-\gamma}, \text{ where}$$

B system bandwidth in MHz
B_1 bandwidth of the fiber at the length L_1 in MHz
L length of the fiber in km
L_1 length of the fiber, usually 1 km, at bandwidth B_1
γ γ factor.

The γ factor is equal to 1 at the beginning of the optical fiber and for long lengths it reaches 0.5. In the intermediate range it varies from 0.5 to 1. It is only possible to determine the γ factor experimentally by measuring the bandwidth of the fiber as a function of the length L. For this purpose the optical fiber must be cut in pieces, which is usually not practical in actual use. It is therefore problematic to convert a measured bandwidth of an optical fiber to another length and at present it is not satisfactorily explained, neither experimentally nor theoretically. This fact, however, does not affect the function of a properly designed optical cable plant.

Therefore in the following only the determination of the bandwidth B of an optical fiber and the magnitude of the transfer function $H(f_m)$ will be discussed. Two methods are available for this purpose:

measurement in the frequency domain and
measurement in the time domain.

Measurement in the Frequency Domain

According to the definition of the transfer function, the amplitude of the light power of a transmitter is modulated with continually increasing frequency f_m, and at the receiver end of the optical fiber the power is determined for this measurement in the frequency range or frequency domain. Together with the equations mentioned above, the magnitude of the transfer function and therefore the bandwidth can be obtained. If a transmitter is available whose amplitude of light power $P_1(f_m)$ is independent of the frequency of modulation f_m, then the bandwidth B can be read directly as that frequency of the transmitter at which the amplitude of the light power $P_2(f_m)$ at the receiver has decreased to half its value at zero frequency. Hence this measurement method requires little time and is useful for measuring the bandwidth of fibers in installed optical cables, because few technical devices are needed.

Measurement in the Time Domain

This measurement is used to study the pulse broadening in time caused by the dispersion effects in the optical fiber. Thereby a short light pulse (typically of 100 ps duration) is launched into the fiber under test. Due to modal and material dispersion the input pulse broadens as it propagates in the optical fiber. The resulting output pulse enters a photo diode in the receiver; there it is amplified and transmitted to the input of a sampling oscilloscope. This measurement must be repeated on a short (approximately 2 m) optical fiber with few perturbations to determine the input pulse.

By means of the recorded data for the input pulse $g_1(t)$ and the output pulse $g_s(t)$ it is possible to calculate by integration the corresponding full root mean square (FRMS) pulse widths T_1 and T_2. From that the FRMS pulse broadening is calculated

$$\Delta T_{\mathrm{FRMS}} = \sqrt{T_2^2 - T_1^2}$$

and an approximate value for the bandwidth

$$B \approx \frac{0.375}{\Delta T_{\mathrm{FRMS}}} .$$

Because this approximation assumes a Gaussian-shaped pulse (therefore the factor 0.375) and thereby ignores the actual structure of the pulse except for its FRMS width, a *Fourier transform* of the pulses from the time domain into the frequency domain must be performed to determine the bandwidth more precisely. In the process integration is used to calculate the magnitude and phase of the transfer function corresponding to these input and output pulses.

In contrast to the measurement of the amplitude in the frequency domain, this also gives the phase of the transfer function, which allows statements about the symmetry of the pulse shapes. The bandwidth is derived from the magnitude of the transfer function; to be precise, it is the frequency at which the magnitude falls to half of what it was at zero frequency. A typical measurement record can be seen in Figure 5.8.

The bandwidth measurement technique is specified in European standard EN 188 000 (national German standard VDE 0888 Part 101) and internationally in IEC 60793-1-4 methods C2A / C2B.

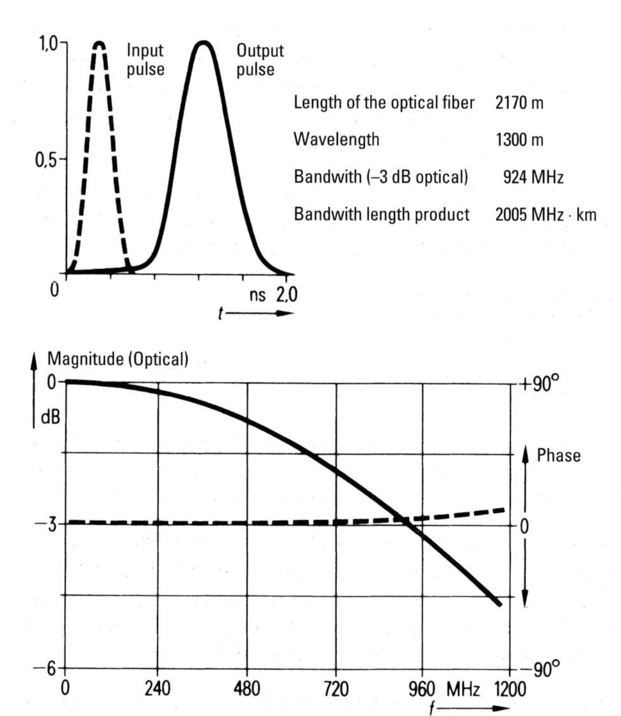

Figure 5.8
Measurement record of a bandwidth measurement – measurement in the time domain
Top: Normalized time pulse
Bottom: Transfer function

5.4 Chromatic Dispersion

Light pulses propagate in an optical fiber at the group velocity

$$c_g = \frac{c}{n_g}$$

where n_g represents the group refractive index of the core glass, which is dependent on the wavelength λ. To traverse through an optical fiber with the length L a light pulse requires a group delay time

$$t_g = \frac{L}{c_g} = \frac{L}{c} \cdot n_g.$$

Hence, due to the group refractive index n_g, the group delay time is dependent on the wavelength λ. All light sources for an optical fiber emit light not only in one single wavelength λ but rather in a *spectral width* $\Delta\lambda$ distributed around the wavelength λ. Therefore the individual light portions within $\Delta\lambda$ propagate at different velocities and have different delay times.

A measure for the variations in the group refractive index n_g at different wavelengths is the *material dispersion* M_0, which is calculated from the first derivative of the group refractive index with respect to the wavelength:

$$M_0(\lambda) = \frac{1}{c} \cdot \frac{\mathrm{d}n_g(\lambda)}{\mathrm{d}\lambda} = \frac{1}{L} \cdot \frac{\mathrm{d}t_g(\lambda)}{d\lambda}.$$

The material dispersion M_0 is usually given in units of $\frac{\text{ps}}{\text{nm} \cdot \text{km}}$.

Because in fused silica glass the group refractive index n_g has a minimum at about the wavelength of 1300 nm, the derivative at this point becomes zero and the material dispersion $M_0(\lambda)$ at this wavelength is infinitely small.

The material dispersion is a material-dependent quantity. By means of the dopants in the core glass it can be varied within certain limits and thereby the zero point can be influenced. It is effective in all optical fibers, both single-mode and multimode. However, for multimode fibers in the vicinity of the zero point its magnitude is greatly exceeded by the modal dispersion.

There is one more dispersion effect to be considered, the waveguide dispersion, which is particularly important in single-mode fibers. It is caused by the wavelength dependence of the distribution of the light of the fundamental mode in the core and cladding glass and therefore by the refractive index difference $\Delta = \Delta(\lambda)$.

The higher the wavelength, the more the fundamental mode LP_{01} widens from the core into the cladding glass (Figure 5.9). This means that an increasing portion of light of the fundamental mode is guided in the cladding with its comparatively low refractive index relative to the core. Therefore the fundamental mode propagates faster. Within the spectral width $\Delta\lambda$ delay time differences develop. The actual velocity of propagation of the fundamental mode is uniform in the core and cladding glass, i.e. an average value for it can be calculated from the velocity of propagation in the core and cladding glass.

The sum of the two dispersions (material dispersion and waveguide dispersion) is called the *chromatic dispersion* $M(\lambda)$:

$$M(\lambda) = M_0(\lambda) + M_1(\lambda).$$

The wavelength λ_0, at which the chromatic dispersion disappears, is called the *zero dispersion wavelength.* Figure 5.10 shows the dispersion curves of an optical fiber with shifted zero dispersion wavelength at 1550 nm.

A light pulse launched into a single-mode fiber from a transmitter having a spectral width (full width half maximum, FWHM) $\Delta\lambda$ or a full root mean square (FRMS) width

($\Delta\lambda_{\mathrm{FRMS}} = \frac{1}{\sqrt{\ln 4}}\,\Delta\lambda \approx 0.85 \cdot \Delta\lambda$ for a Gaussian-shaped transmitter spectrum) is modified in time by the chromatic dispersion $M(\lambda)$. For an FRMS pulse width T_1 at the beginning and T_2 after the length L, the FRMS pulse broadening ΔT_{FRMS} is calculated as

$$\Delta T_{\mathrm{FRMS}} = \sqrt{T_2^2 - T_1^2} = M(\lambda) \cdot \Delta\lambda_{\mathrm{FRMS}} \cdot L.$$

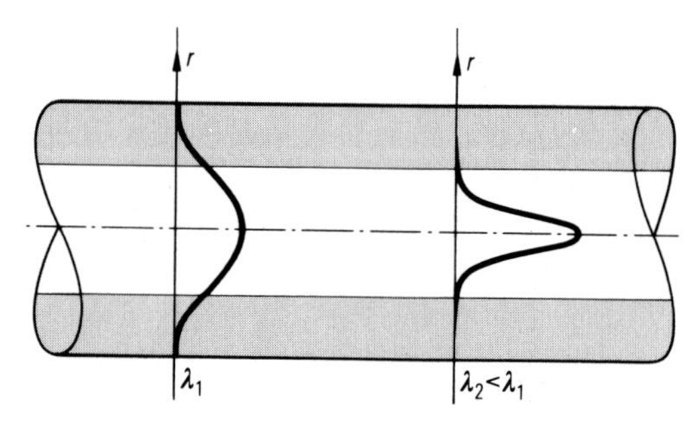

Figure 5.9
Energy distribution of the fundamental mode at two different wavelengths

The pulse broadening increases in proportion to the length L of the optical fiber. It is also proportional to the spectral width $\Delta\lambda$ of the transmitter.

Typical values for a laser diode (LD) are $\Delta\lambda$ = 3 to 5 nm and for a light emitting diode (LED) $\Delta\lambda$ = 40 to 70 nm (at 850 nm) and $\Delta\lambda$ = 120 to 150 nm (at 1300 nm).

The above equation applies for Gaussian-shaped pulses and spectra, whereby the center wavelength λ_m of the transmitter may not lie close to the wavelength λ_0, at which the chromatic dispersion $M(\lambda)$ is zero. In that case the general equation for the FRMS pulse broadening ΔT_{FRMS} must be used:

$$\Delta T_{FRMS} = \Delta\lambda_{FRMS} \cdot L \cdot \sqrt{M(\lambda_m)^2 + M'(\lambda_0)^2 \cdot \frac{\Delta\lambda_{FRMS}^2}{8}} \; .$$

where

$M(\lambda_m)$ dispersion at the center wavelength λ_m of the transmitter in $\dfrac{\text{ps}}{\text{nm} \cdot \text{km}}$

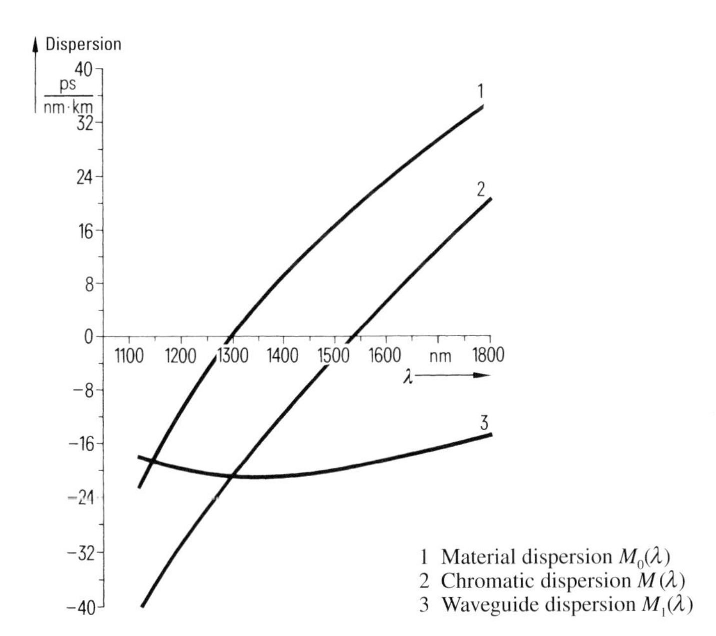

Figure 5.10
Dispersion curves of a single-mode fiber

$M'(\lambda_0)$ slope of the dispersion at the zero dispersion wavelength λ_0 in

$$\frac{\text{ps}}{\text{nm}^2 \cdot \text{km}}$$

$\Delta\lambda_{\text{FRMS}}$ FRMS width of the transmitter in nm.

The bandwidth B of a single-mode fiber can be calculated as a function of the center wavelength λ_m for a Gaussian-shaped transmitter spectrum and an FRMS pulse broadening ΔT_{FRMS}:

$$B = \frac{\sqrt{\ln 4}}{\pi} \cdot \frac{1}{\Delta T_{\text{FRMS}}} = \frac{\ln 4}{\pi} \cdot \frac{1}{\Delta T} \approx \frac{0{,}375}{\Delta T_{\text{FRMS}}} \approx \frac{0{,}441}{\Delta T} \,.$$

Figure 5.11 shows the bandwidth curves for a typical single-mode fiber and different spectral widths $\Delta\lambda$.

5.4.1 Measurement Methods

When measuring the chromatic dispersion $M(\lambda)$ as a function of the wavelength λ, usually a transmitter with variable wavelength or several transmitters with different but fixed wavelengths are employed.

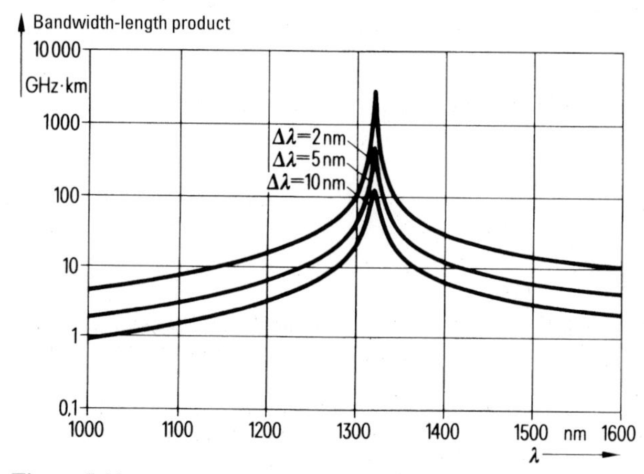

Figure 5.11
Bandwidth length product of single-mode fibers dependent on the spectral width $\Delta\lambda$ of the transmitter

A neodymium-yttrium aluminium garnet (Nd-YAG) laser is suitable for use as a high power transmitter for the wavelength range from 1064 to about 1800 nm. It generates a *Raman scattering*[1)] of the light in a quite heavily doped single-mode fiber several hundred meters long. With a monochromator the desired measuring wavelength is filtered out and launched into the fiber under test. At the end of the optical fiber a receiver and an oscilloscope are used to measure the absolute group delay time t_g for the length L of the optical fiber. By carrying out several measurements the group delay time t_g can be obtained as a function of the wavelength λ (Figure 5.12).

The chromatic dispersion $M(\lambda)$ is derived from the group delay curve by calculating the slope of this curve and dividing it by the length L of the optical fiber. Figure 5.13 shows the corresponding curves of the chromatic dispersion.

[1)] Nonlinear optical effect due to intense light radiation with a wide continuous spectrum.

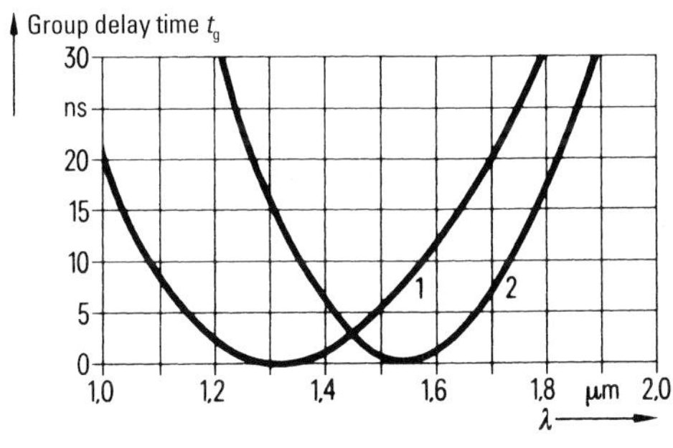

1 Single-mode fiber with zero dispersion wavelength at 1.3 µm
2 Single-mode fiber with zero dispersion wavelength at 1.55 µm

Figure 5.12
Group delay time of single-mode fibers as a function of the wavelength

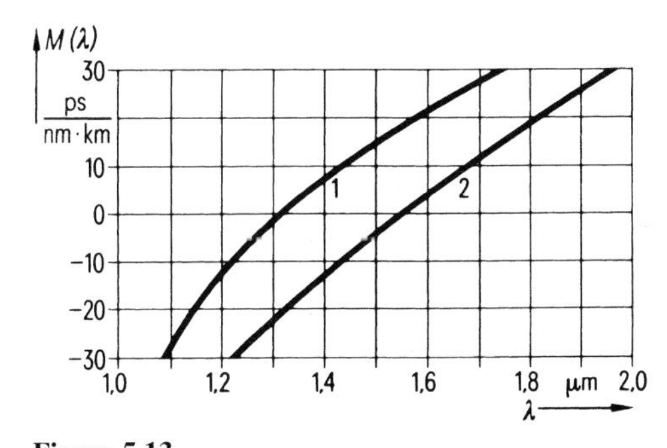

1 Single-mode fiber with zero dispersion wavelength at 1.3 µm
2 Single-mode fiber with zero dispersion wavelength at 1.55 µm

Figure 5.13
Chromatic dispersion of single-mode fibers as a function of the wavelength

In addition to this dispersion measurement by means of the Raman scattering it is also possible to use several laser diodes with wavelengths close to the zero dispersion wavelength. This permits the measurement of delay times t_g for these specific wavelengths. By interpolation and curve fitting these can be used to calculate the group delay time curve. The further evaluation is carried out corresponding to the Raman method.

Such a laser diode setup is especially handy and useful in the field, if desired. Due to the predetermined wavelengths, however, it is questionable whether it can be used to define the chromatic dispersion of multistep single-mode profiles with sufficient accuracy.

The dispersion measurement technique is specified in European standard EN 188 000 (national German standard VDE 0888 Part 101) and internationally in IEC 60793-1-4 methods C5A / C5B.

5.5 Cut-Off Wavelength

Depending on wavelength, either one mode or several modes propagate in an optical waveguide. For the operation of an optical waveguide it is important to determine the wavelength at which only the fundamental mode LP_{01} is guided in the waveguide in question. The cut-off wavelength of an optical waveguide is the term used for the lowest operational wavelength, beyond which only the fundamental mode is propagative. Above this cut-off wavelength λ_c, the next higher mode LP_{11} disappears. Below λ_c further modes are added at lower wavelengths. The optical waveguide is multimode at wavelengths lower than λ_c and single-mode at higher wavelengths.

5.5.1 Measurement Methods

Bending Method

A method frequently used to determine the cut-off wavelength is the bending method. It involves measuring the spectral light power, i.e. the light power as a function of the wavelength, in a 2 m long piece of the optical fiber under test. First it is measured with the fiber in as straight a position as possible (bending radius larger than 140 mm); then the fiber is wrapped once around a mandrel with a diameter of 30 mm and measured. The attenuation due to the bending is calculated as a function of the wavelength and plotted on a graph (Figure 5.14). This gives several well-defined maximum points for the

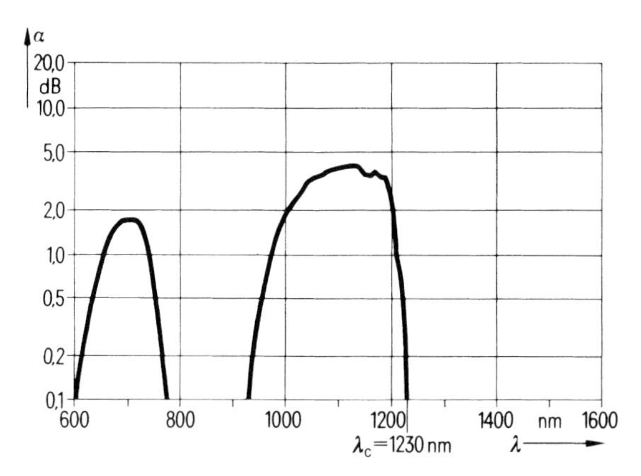

Figure 5.14
Determination of the cut-off wavelength by the bending method

attenuation with steep slopes toward higher wavelengths. These slopes mark the cut-off wavelengths of the lowest modes. The cut-off wavelength λ_c of the LP_{11} mode is then the wavelength at which the downward slope of the maximum with the highest wavelength passes through the attenuation value of 0.1 dB.

The value λ_c measured by this method is a characteristic not only of the design of the optical fiber but depends also on the length tested. The shorter the optical fiber, the higher will be its cut-off wavelength. The theoretical cut-off wavelength for a piece of optical fiber a few millimeters in length is about 100 nm longer than for a piece 2 m long.

The cut-off wavelength measurement technique is specified in European standard EN 188 000 (national German standard VDE 0888 Part 101) and internationally in IEC 60793-1-4 method C7A.

5.6 Mode Field Diameter

The light distribution of the fundamental mode in a single-mode fiber is an important factor in the evaluation of the launch, bending and joint losses. The mode field radius w_0 or the *mode field diameter* $2w_0$ have been defined to describe this distribution (Figure 5.15).

The mode field radius w_0 is the radius at which the radial field amplitude decreases to the $1/e$-th value (≈ 37 per cent because $e \approx 2.71828$) of its maximum at the optical fiber axis ($r = 0$). The mode field diameter is dependent on the

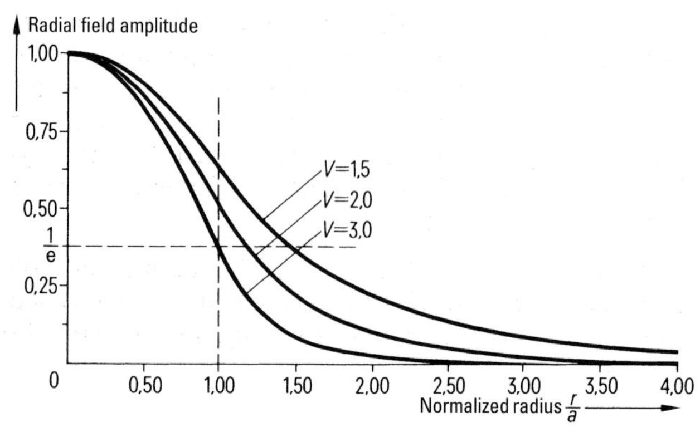

Figure 5.15
Radial dependence of the field amplitude of the fundamental mode

wavelength: it increases as the wavelength increases. The mode field radius w_0 relative to the core radius a is only a function of the V number, which in turn is dependent on the wavelength λ and the numerical aperture NA as follows:

$$V = \pi \cdot \frac{2 \cdot a}{\lambda} \cdot NA.$$

There is a simple approximation for a single-mode step index fiber, which in the range of the V number $1.6 < V < 2.6$ quite accurately describes the relation between the mode field radius w_0, the core radius a and the V number:

$$w_0 = \frac{2.6}{V} \cdot a.$$

The above V number range corresponds to the wavelength range between 1150 and 1875 nm and therefore covers the usual operating wavelengths of 1300 and 1550 nm.

5.6.1 Measurement Methods

Transverse Offset Method

There are different methods for the measurement of the mode field diameter. The transverse offset method is the one usually used. It involves coupling a 2 m long piece of the optical fiber under test to a spectral attenuation measuring setup, which is used to measure the light power as a function of the wavelength. Then the test sample is broken in the middle and the two ends are placed in holders of a micropositioning unit so that they lie closely opposite to each other (gap less than one core diameter). The positioning unit is used to move one fiber end so that a maximum light transmission is measured at the receiver of the spectral attenuation measurement setup. Then the light transmission is measured as a function of the radial transverse offset of the optical fiber ends by moving one optical fiber end in equal steps and registering the reduction in light power. The $1/e$ diameter of the resulting curve gives the mode field diameter $2w_0$ (Figure 5.16).

The measuring wavelength for the mode field diameter has been set at 1300 nm. If the mode field diameter or mode field radius is measured as a function of the wavelength, then a curve results, such as the one shown in Figure 5.17

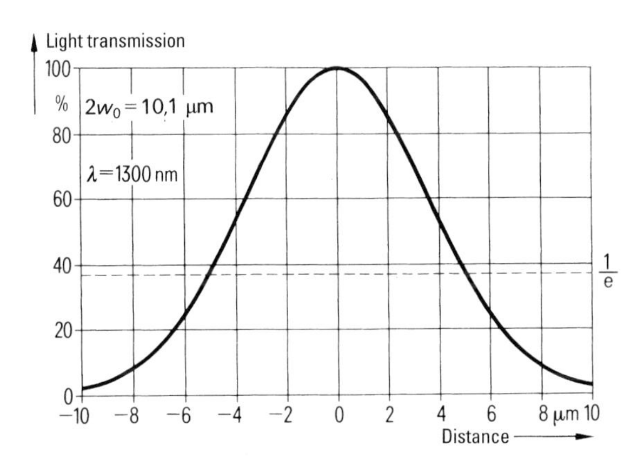

Figure 5.16 Determination of the mode field diameter by the transverse offset method

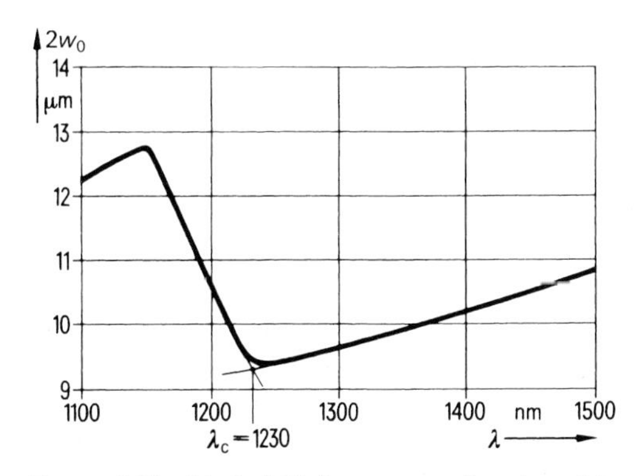

Figure 5.17 Mode field diameter as a function of the wavelength

for a typical single-mode fiber. The steep decrease of the mode field diameter $2w_0$ at 1200 nm indicates the cut-off wavelength λ_c of the LP_{11} mode.

The mode field diameter measurement technique is specified in European standard EN 188 000 (national German standard VDE 0888 Part 101) and internationally in IEC 60793-1-4 methods C9A / C9B.

5.7 Near Field Distribution and Geometric Measurements

The near field distribution of an optical fiber is understood to be the distribution of the light power density over the cross-sectional area at the end of an optical fiber.

5.7.1 Measurement Methods

Refractive Index Profile

Because the distribution of optical power in an multimode fiber is proportional to the distribution of the refractive index, the near field distribution can be used to determine the refractive index profile of an optical fiber. It is measured by using a white source, e.g. a halogen lamp, for full flood launch into a short piece of optical fiber so that all modes are excited equally. Cladding modes are removed by a mode stripper. At the other end of the fiber the light intensity $I_N(r)$ over the cross-sectional area of the optical fiber is observed through a microscope and detector (Figure 5.18).

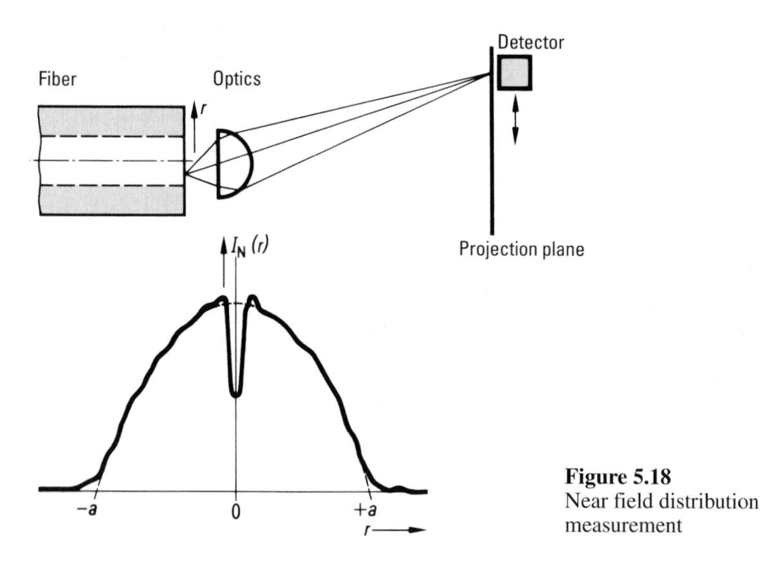

Figure 5.18
Near field distribution measurement

In addition to the refractive index profile, this configuration can also be used to determine the *core* and *cladding diameter*, the *concentricity* of the core in the cladding and the *noncircularity* of the core and cladding (Figure 5.19).

Core diameter:

$$d = \frac{d_{max} + d_{min}}{2}$$

Cladding diameter:

$$D = \frac{D_{max} + D_{min}}{2}$$

Core noncircularity:

$$e = \frac{(d_{max} - d_{min})}{d_0} \cdot 100\ \%$$

Core ellipticity:

$$\varepsilon = \sqrt{1 - \left(\frac{d_{min}}{d_{max}}\right)^2}.$$

Cladding noncircularity:

$$E = \frac{(D_{max} - D_{min})}{D_0} \cdot 100\%$$

Concentricity:

$$c = \left(\frac{x}{d}\right) \cdot 100\%$$

d core diameter
D cladding diameter
e core noncircularity
d_0 core diameter if circular
E cladding noncircularity
D_0 cladding diameter if circular
c concentricity
x deviation from core center
ε core ellipticity

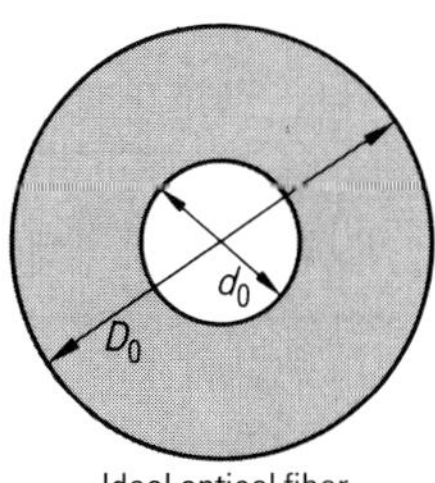

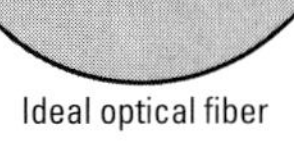

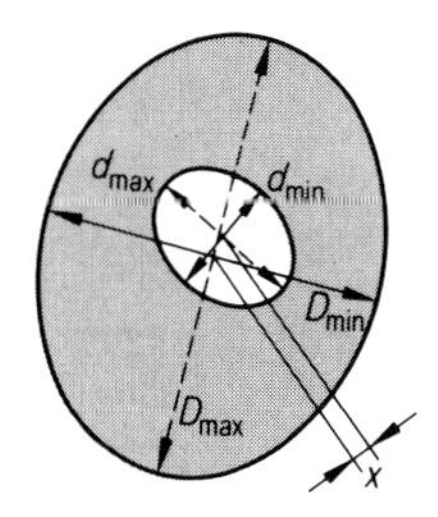

Figure 5.19 Concentricity and noncircularity of core and cladding

Four-Circle Method

Another method of checking the geometric dimensions is the "four-circle method" (Figure 5.20). In this method, by trial and error, four concentric circles – two for the tolerance range of the core diameter and two for that of the cladding diameter – must be juxtaposed to correspond to the actual dimensions of the optical fiber cross-sectional area.

During the measurement of the near field distribution the full flood launch also launches leaky modes into the short piece of optical fiber, so that the measured refractive index profile does not correspond exactly to reality. Use of the *refracted near field method* avoids this problem. In it the light which does *not* propagate through the optical fiber is measured (Figure 5.21).

If the refractive index of the cladding glass is known (it is usually of pure SiO_2), then this method makes it possible to measure the absolute refractive index profile with great precision. An example for a single-mode fiber with multistep core is graphed in Figure 5.22.

The four-circle measurement technique is specified in European standard EN 188 000 (national German standard VDE 0888 Part 101) method 103 and the refractive index profile measurement technique internationally in IEC 60793-1-2 method A1.

5.8 Far Field Distribution

The far field distribution of an optical fiber is understood to be the distribution of the light power $I_F(\Theta)$ radiated from the end of an optical fiber over the angle Θ relative to the axis of the fiber.

5.8.1 Measurement Methods

A configuration similar to the setup for the measurement of the near field distribution can be used to measure the far field distribution. Only the microscope is left out to observe the end face of the optical fiber and the receiver must be swivel mounted for circular movement about the output end of the optical fiber. The light power is then measured as a function of the angle Θ (Figure 5.23).

The measurements usually result in a Gaussian-shaped intensity distribution of the far field. Using the maximum radiation angle Θ_{max}, the numerical aperture NA of the optical fiber can be calculated from these results:

$$NA = \sin \Theta_{max}.$$

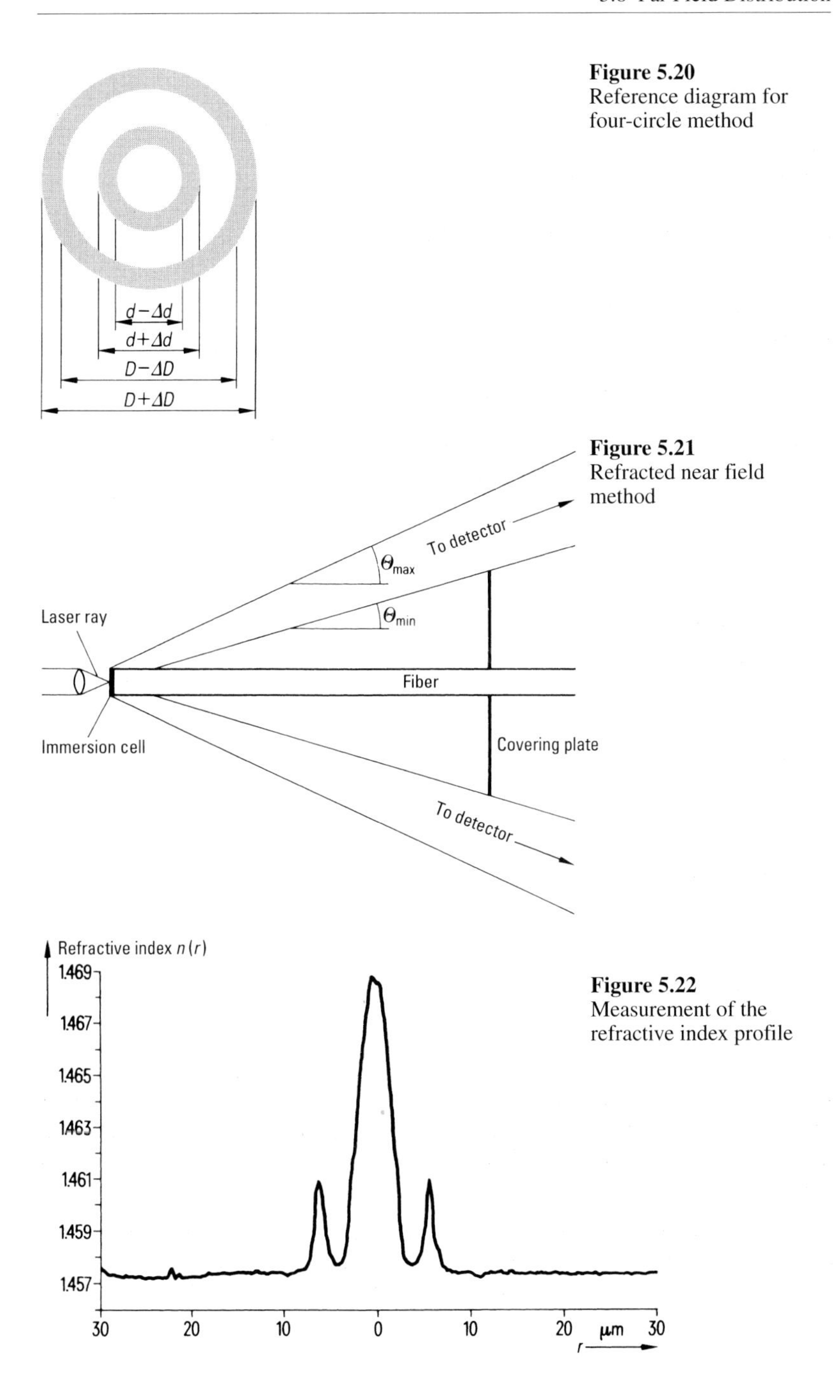

Figure 5.20
Reference diagram for four-circle method

Figure 5.21
Refracted near field method

Figure 5.22
Measurement of the refractive index profile

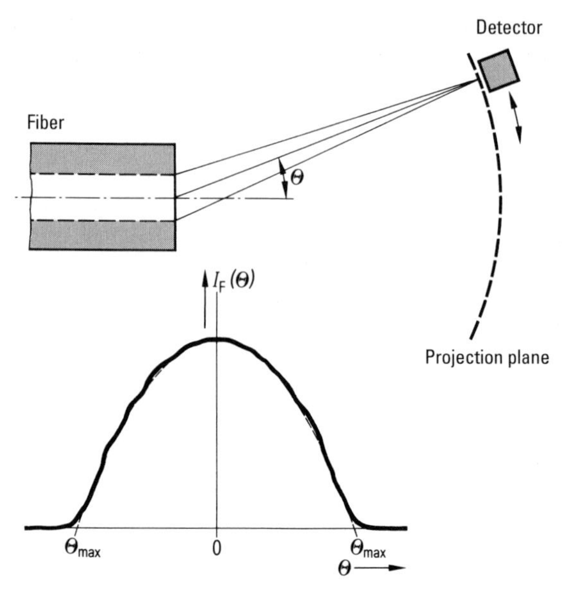

Figure 5.23 Far field distribution measurement

It is difficult to measure Θ_{max} due to the low intensity at this maximum angle of radiation. To determine this angle, either a tangent is laid at the intensity curve and its point of intersection with the abscissa is defined as Θ_{max} or the angle at 10 per cent intensity is taken to be Θ_{max}.

The numerical aperture measurement technique is specified in European standard EN 188 000 (national German standard VDE 0888 Part 101) and internationally in IEC 60793-1-4 method C6.

5.9 Mechanical Strength

Due to the inhomogeneity of the materials in the glass or perturbations on the surface of the glass, mechanically weak points can be found along an optical fiber. Because these irregularities can only be accounted for in a statistical manner, the mechanical strength of an optical fiber is only predictable with a certain probability. Furthermore, the fracture of an optical fiber at one of these perturbations is a time-dependent process.

A *Weibull distribution* is used to describe mathematically the probability of fracture F in an optical fiber dependent on the length L, the mechanical break stress σ and the time t.

$$F(L, \sigma, t) = 1 - \exp\left\{-\frac{L}{L_0}\cdot\left(\frac{\sigma}{\sigma_0}\right)^a\cdot\left(\frac{t}{t_0}\right)^b\right\}.$$

F probability of fracture

L length of the fiber in m

σ break stress in $\frac{\mathrm{N}}{\mathrm{mm}^2}$

t time in s

L_0 test length in m

σ_0 nominal stress in $\frac{\mathrm{N}}{\mathrm{mm}^2}$

a constant

t_0 nominal duration of test in s

b constant

The quantities L_0, σ_0 and t_0, in addition to the two other constants a and b, must be determined experimentally. For $L = L_0, \sigma = \sigma_0$ and $t = t_0$, it is true that

$$F(L_0, \sigma_0, t_0) = 1 - \frac{1}{e} \approx 0.632 \text{ (nominal value of the distribution).}$$

The quantity σ_0 is called the nominal stress, t_0 the nominal duration of test and L_0 the length being tested. Typical values for the constants a and b are

$a = 3 \pm 1$ and
$b = 0.2 \pm 0.05$.

The test length L_0 is typically 20 m, the nominal duration of test t_0 approximately 1 s and the nominal stress σ_0 approximately 2000 N/mm^2.

Figure 5.24 shows a Weibull distribution without time dependence $t = t_0$; the break stress is plotted on the abscissa in a logarithmic scale and the probability of fracture F is plotted on the ordinate in the scale:

$$\log \ln \frac{1}{1 - F(L, \sigma)}$$

The slope of the straight line through the measured points gives the constant a.

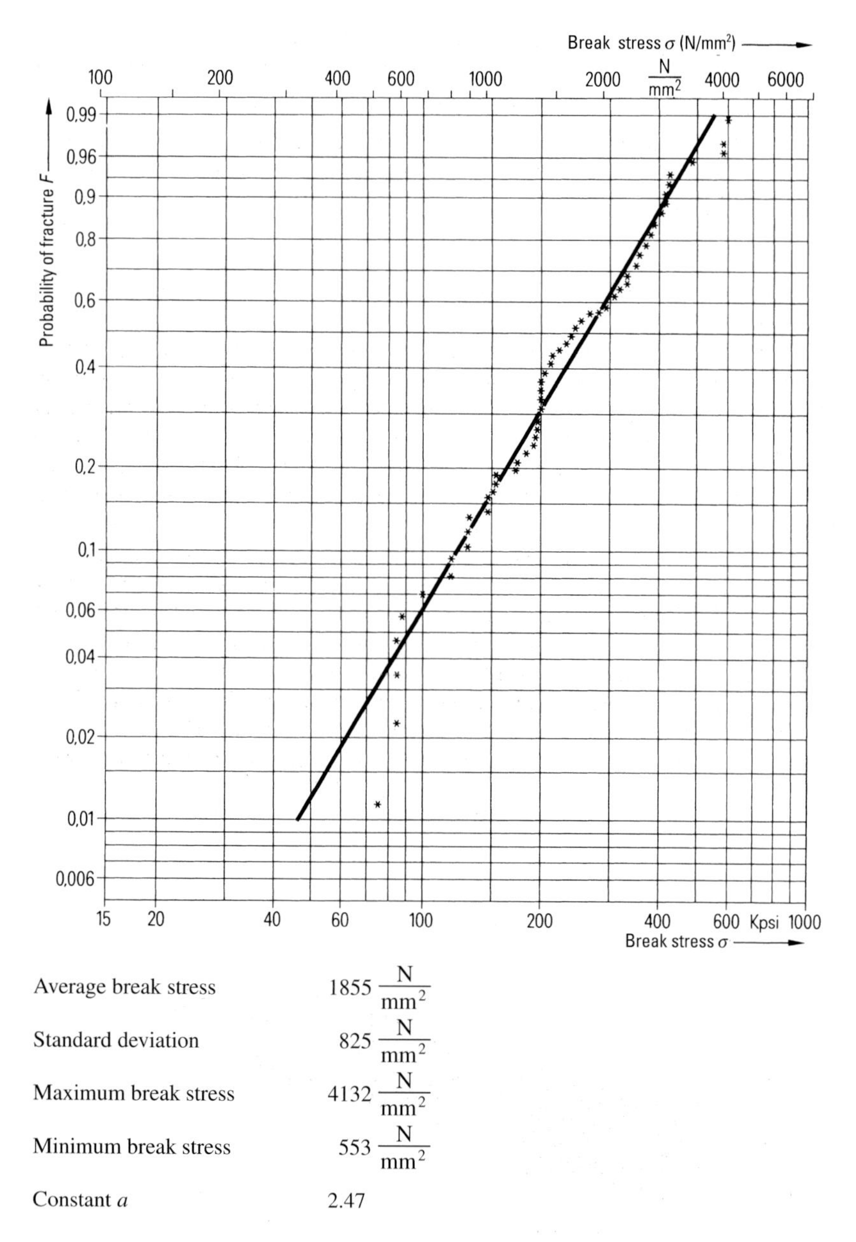

Average break stress	$1855 \frac{N}{mm^2}$
Standard deviation	$825 \frac{N}{mm^2}$
Maximum break stress	$4132 \frac{N}{mm^2}$
Minimum break stress	$553 \frac{N}{mm^2}$
Constant a	2.47

Figure 5.24
Weibull distribution for determination of the mechanical strength of an optical fiber

Please note that during production of optical fibers their mechanical strength is checked in a proof test or screen test. Usually a system of rollers is used to exert a stress of 345 N/mm^2 on the fiber as it runs through them.

The mechanical strength measurement techniques are specified in European standard EN 188 000 (national German standard VDE 0888 Part 101) and internationally in IEC 60793-1-3 method B2.

5.10 Polarization Mode Dispersion

A so-called single-mode fiber is not really single mode (Chapter 4). In fact the fundamental mode in a single-mode is degenerated twice, i.e. it consists of two orthogonally (mutually perpendicular) polarized modes which travel always simultaneously. In an optical fiber with ideal optically circular symmetric core both polarization modes propagate with identical velocities. However, actually manufactured optical fibers exhibit some birefringence due to unavoidable small asymmetries in the core geometry (ellipticity) or internal stresses frozen within the glass during the fiber drawing process. Other sources of birefringence are fiber strain, fiber side pressure, fiber bending. This leads to an anisotropic refractive index and thus different velocities of the two polarization modes. As a consequence light pulses injected into a birefringent fiber are prone to separate as shown schematically in Figure 5.25. The two polarization modes travel with different velocities according to the different refractive indices of the "slow" and the "fast" fiber axis. The separation of the modes (in ps) at the end of the fiber corresponds to the polarization mode dispersion (PMD).

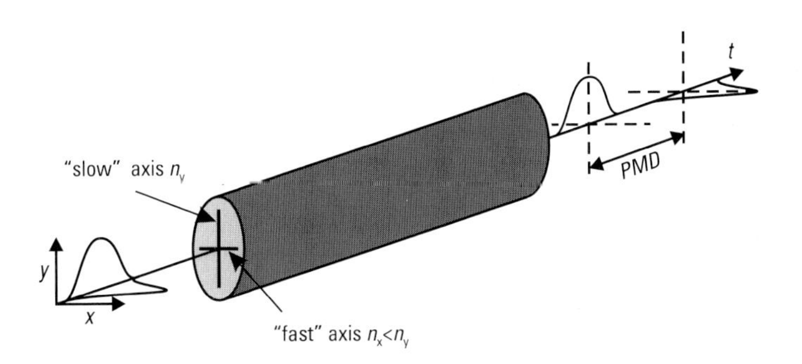

Figure 5.25 Separation of polarization modes in a birefringent optical fiber

Due to a random mode coupling between the two polarization modes energy is partly exchanged between them which leads to the statistical nature of the PMD. For a fiber length L shorter than the so-called coupling length L_c the PMD evolves linearly with L. For a fiber length L larger than the coupling length L_c the PMD increases in proportion to the square-root of the fiber length L.

As mentioned above a core ellipticity ε (Chapter 5.7.1) can induce birefringence B.

$$B = \text{const} \cdot \varepsilon^2 \qquad (\text{for } \varepsilon < 0.3).$$

Especially cabled optical fibers are subject to side pressure which may create birefringence.

$$B = 4n_k^3 \cdot (1 + \mu) \cdot (p_{12} - p_{11}) \cdot Q/(\pi \cdot E \cdot 2 \cdot A).$$

n_k	core refractive index
μ	Poisson's ratio
E	Young's modulus
Q	lateral force
p_{11}, p_{12}	photoelastic constants
A	cross-sectional area of the glass fiber.

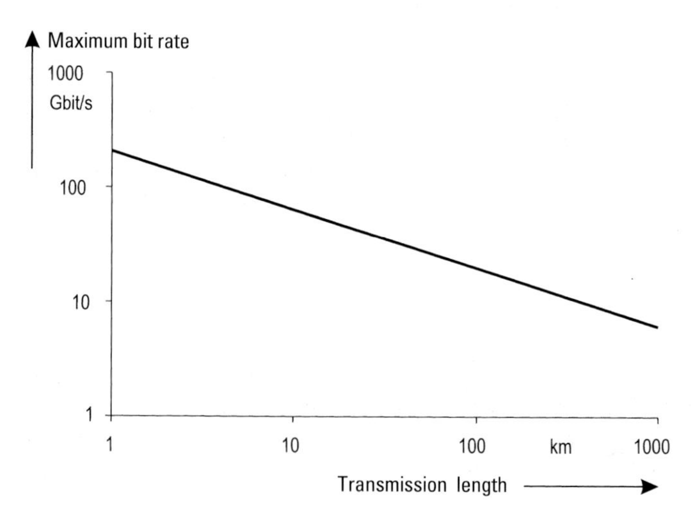

Figure 5.26 Bit rate versus transmission length for PMD below 0.1 of a bit period

Pure fiber bending also creates an asymmetry of the refractive index by an inhomogenious strain distribution across the bent fiber. In the case of a pure bending the birefringence is given by

$$B = n_k^3 \cdot (1 + \mu) \cdot (p_{12} - p_{11}) \cdot (A/R)^2/4,$$

where R is the bending radius (the other quantities being defined above).

Cabling does not significantly influence the PMD. The distribution of the PMD-values measured on optical fibers loosely wound on measurement drums (diameter = 30 cm) is almost identical to fibers which are inserted in buffer tubes stranded in a standard telecommunication cable. It has to be noted, that the majority of the measured PMD values are well below 0.5 ps/$\sqrt{\text{km}}$ which has become an acknowledged international specification.

PMD in optical fibers can influence digital as well as analog optical transmission of information. As already discussed in Chapter 5.4 the chromatic dispersion can be minimized by selecting small spectral width light sources (e.g. DFB-lasers, Chapter 12.1.2) and/or by applying a suitable technique for dispersion compensation (e.g. dispersion compensation modules, Chapter 13.2). PMD, however, cannot be compensated for and therefore ultimately limits the transmission length (the power budget is neglected here). Standards groups dealing with amplified long-distance systems tend to a requirement which keeps mean values of PMD below 0.1 of a bit period. Figure 5.26 demonstrates the consequence of a PMD value of 0.5 ps/$\sqrt{\text{km}}$ for the bit rate and distance relation.

Analog systems sometimes suffer from so-called composite second order (CSO) distortions which lead to visible artifacts on TV screens. This effect is the result of a complex interplay of PMD, polarization dependent loss (PDL) and laser chirp. With a maximum PMD value for optical fibers of 0.5 ps/$\sqrt{\text{km}}$ no signal distortions should occur up to a link length in the order of 250 km for a CSO power level of about –65 dBm and 2% PDL.

5.10.1 Measurement Methods

Polarization mode dispersion can be measured in the time as well in the frequency domain. The *interferometric method* (time domain), the *state of polarization analysis* (frequency domain) and the *fixed analyzer method* (frequency domain) are being used.

Interferometric Method

The experimental set-up is based on a Michelson or a Mach-Zehnder interferometer (Figure 5.27). Linearly polarized light from an LED with a typical spectral width of 80 nm is launched into the test fiber. The fiber end is connected to a fiber coupler. One branch of the interferometer contains a variable air gap. The interference pattern at the output of a second fiber coupler (combiner) is recorded as a function of the delay time introduced by the air gap. The maximum measurable delay (typically up to 150 ps) depends only on the total length of the air gap. The minimum measurable PMD is limited by the coherence time t_c of the LED. Thus $PMD_{min} = t_c = 0.1$ ps. Typical measurement curves of optical fibers with high and low birefringence are shown in Figure 5.28. The "distance" between the outer peaks to the inner autocorrelation peak is the PMD in ps. In the case of strong mode coupling, the second moment of the interference pattern yields the PMD.

State of Polarization Analysis

A measurement of the PMD in the frequency domain requires a tunable laser, a polarization controller and a polarimeter. The determination of the PMD at a wavelength λ_0 requires the measurement of the polarization state of the out-

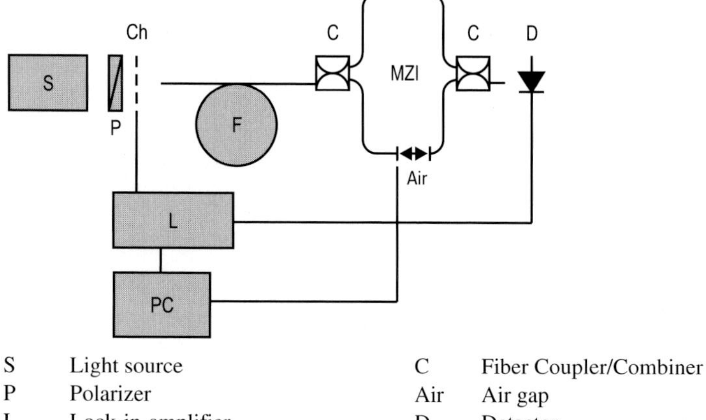

S	Light source	C	Fiber Coupler/Combiner
P	Polarizer	Air	Air gap
L	Lock-in-amplifier	D	Detector
PC	Computer	Ch	Chopper
MZI	Mach-Zehnder-Interferometer	F	Fiber under test

Figure 5.27
PMD measurement set-up using a Mach-Zehnder interferometer

coming light (by using the so called Jones-matrix-eigenanalysis) for 3 different polarization states of the incoming light at least at 2 adjacent wavelengths $\lambda_0 - \Delta\lambda$ and $\lambda_0 + \Delta\lambda$. These wavelengths have to be close enough to λ_0 to avoid 2π ambiguity in the determination of the polarization state changes. Therefore $\Delta\lambda$ has to fulfill the relation $\Delta\lambda < \lambda_0^2 / (c \cdot t_c \cdot 2)$. Also the coherence time t_c

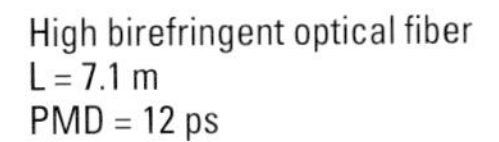

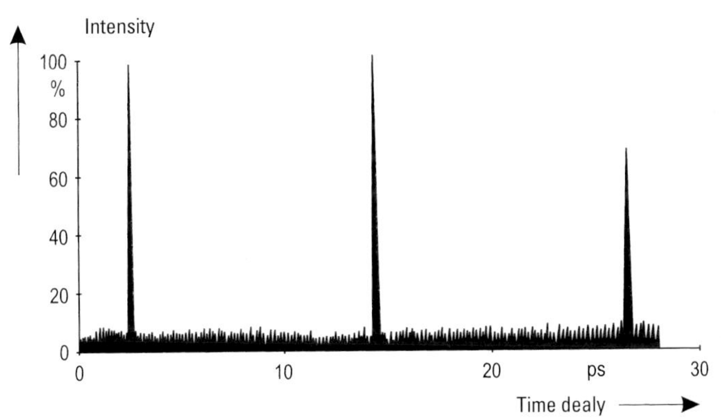

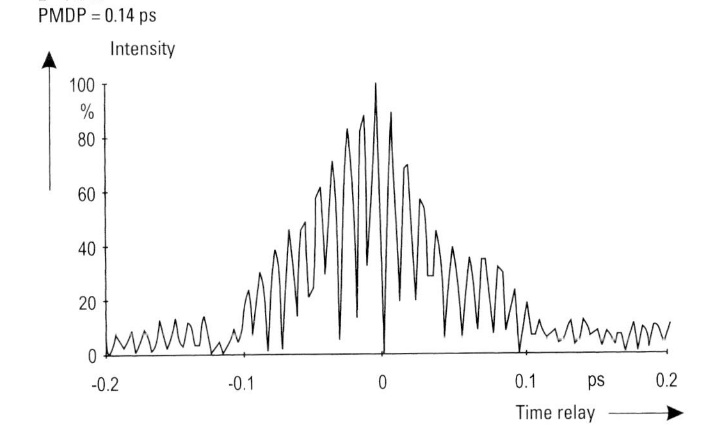

Figure 5.28
Interference patterns for high (low mode coupling) and low birefringent (strong mode coupling) optical fibers

of the laser must be larger than the PMD. A maximum coherence time t_c of e.g. 50 ps requires a spectral width of less than 0.1 nm for the laser. To get reproducible and stable PMD values it is necessary to average over many wavelengths and "time".

Fixed Analyzer Method

The light of a polarized, tunable laser with narrow spectral width is launched into the test fiber. The light at the fiber end passes through an analyzer (fixed) and the light intensity is recorded with varying wavelength (Figure 5.29). In case of a high birefringent fiber the intensity varies perodically with the wavelength. The phase difference Φ between the two polarization modes is given by:

$$\Phi = 2 \cdot \pi \cdot (n_y - n_x) \cdot L/\lambda,$$

where

λ wavelength
L length of optical fiber under test
n_x, n_y refractive indices of the "slow" and "fast" axis.

The PMD follows from:

$$\text{PMD} = N \cdot k \cdot \lambda_1 \cdot \lambda_2/(c \cdot (\lambda_2 - \lambda_1)),$$

where

N number of oscillations
c speed of light
k = 1 for low mode coupling, = 0.8 for strong mode coupling
λ_1, λ_2 start and end wavelength of the wavelength scan.

If low birefringent fibers (strong mode coupling) are investigated, the beat-pattern consists of many frequencies (Figure 5.29). Fourier transform (FT) analysis leads to a pattern which is similar to that obtained by the interferometric method described above. A Gaussian fit or the calculation of the second moment yields the PMD. As this measurement is done in the frequency domain the coherence length of the laser limits the maximum measurable PMD. Additionally the number of points used to compute the FT has to be large enough to avoid any high frequency cut-off.

The PMD measurement procedures are specified in the American standard EIA/TIA-455-124 (PMD measurement for single-mode optical fibers by interferometric method), EIA/TIA-455-122 (PMD measurement of single-mode optical fibers by Jones matrix eigenanalysis) and EIA/TIA-455-113 (PMD measurement of single-mode optical fibers by wavelength scanning) and internationally in ITU-T G.650. European standards are under discussion.

5.11 Return Loss

The return loss (RL) of an optical component is the ratio of the incident power P_{in} to the reflected optical power P_{ref} in units of dB.

$RL = 10 \cdot \log(P_{in}/P_{ref})$.

Because of the sensitivity of lasers to their own light being reflected from optical components back to them, the return loss of such optical components (e.g. connectors, Chapter 10.3.4) is gaining attention. Excellent connectors, which are based on physical contact, exhibit a return loss of approximately 60 dB.

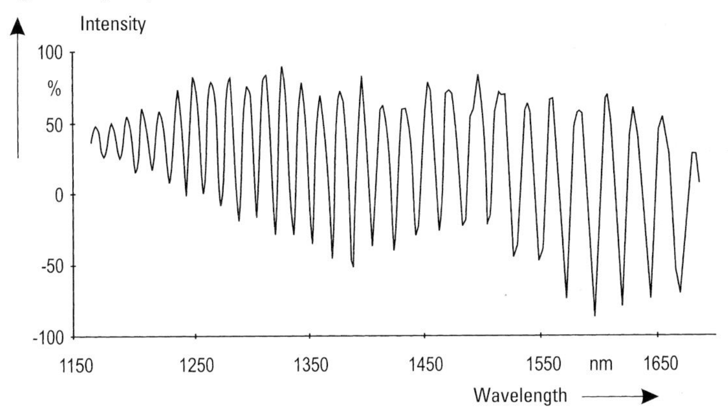

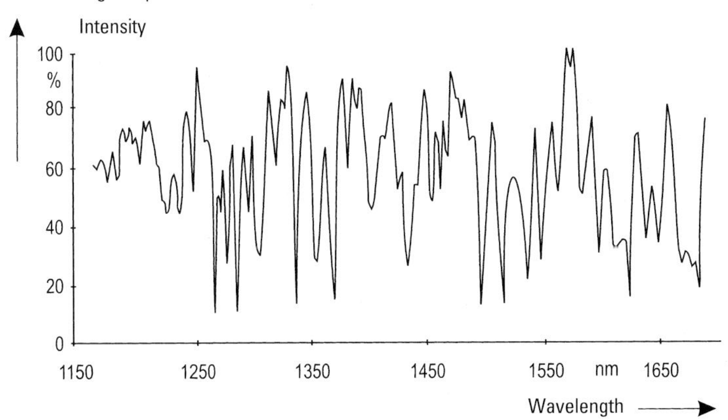

Figure 5.29 Beat patterns of high and low birefringent fibers

5.11.1 Measurement Methods

Return Loss Measurement with an OTDR

In Figure 5.30 an optical time domain reflectometer (OTDR) trace is presented schematically (Chapter 5.2.1). The peak in the curve is due to a reflection caused for instance by a connector. If the height of the peak (in dB) is labeled A, then A is defined as follows:

$$A = 5 \cdot \log\,[(P_{ref} + w \cdot S \cdot P_{in})/(w \cdot S \cdot P_{in})].$$

P_{ref} reflected power
P_{in} incoming power at position L
w pulse width
S capture fraction

The term $w \cdot S \cdot P_{in}$ determines the backscattered power. Basic mathematical transformations lead to:

$$RL = -10 \cdot [\log(w \cdot S) + \log(10^{A/5} - 1)] = K + \log(10^{A/5} - 1).$$

Thus the RL can be evaluated by simply measuring the height of the peak A if the constant K is known. By performing a calibration measurement, e.g. by cutting the optical fiber and creating a glass/air transition with a well-defined reflection of 14.5 dB, K can be determined.

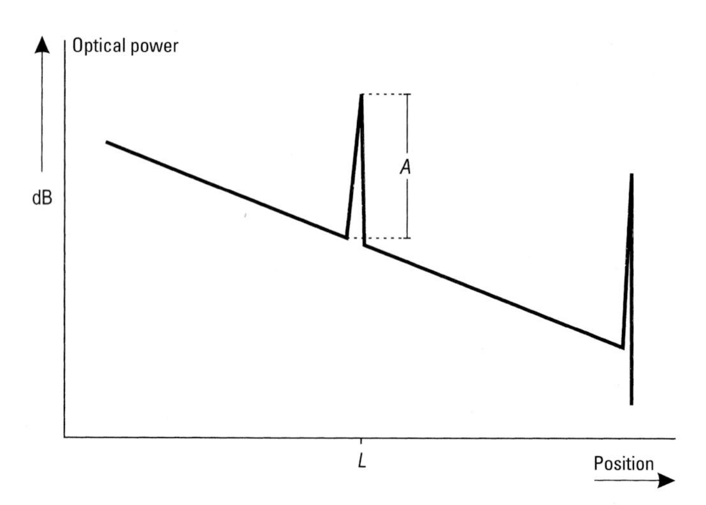

Figure 5.30 OTDR trace with a reflection at position L

Return Loss Measurement with a Fused Fiber Coupler

This method is based on a fused fiber coupler together with a continuous light source and an optical power meter (Figure 5.31). To increase the dynamic range of the measurement a laser source is recommended. In the following the *RL* measurement of a single-mode connector pair as the device under test (DUT) is given as an example.

In this set-up, the DUT is connected to the coupler with a first jumper cable, because connector 4 (the termination of the coupler) should not be part of the DUT. The procedure consists of three steps:

Step A: Calibration of the set-up for –14.5 dB return loss with connector 6 open (glass/air transition). The power meter reads *P1*.

Step B: Measurement of the unwanted reflections from connector pair 4/5 with connector 6 immersed in oil, in order to avoid reflections from the fiber end. The power meter reads *P2*.

Step C: Measurement of the DUT (connector pair 6/7) plus the unwanted reflections. The end of the second jumper cable is now immersed in oil. The power meter reads *P3*.

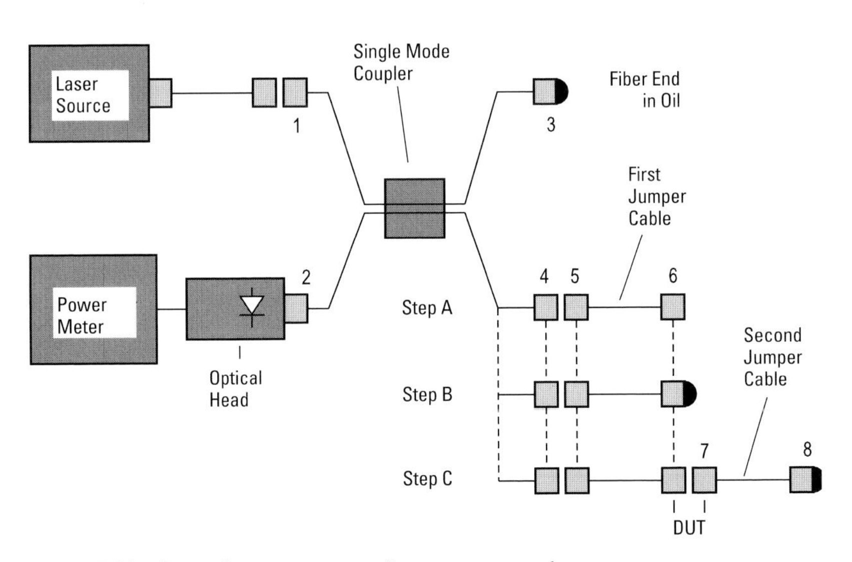

Figure 5.31 Set-up for measurement of connector return loss

The *RL* can be calculated as follows:

From the calibration (Step A) the input power P_{in} is deduced:

$P_{in} = 10^{1.45} \cdot P1$.

Steps B and C yield the reflected power $P_{ref} = P3 - P2$ and therefore:

$RL = 10 \cdot \log[(10^{1.45} \cdot P1)/(P3 - P2)]$.

The return loss measurement with a fused coupler is specified in the American standard EIA/TIA-455-107.

6 Optical Fiber Construction

An optical fiber is composed of two basic parts:

the core and cladding made of an optically transparent material (e.g. fused silica glass) and the coating.

The core is in the center of the optical fiber and is used to guide the light. This wave guidance is only possible in the core because the refractive index of the core n_1 is larger than the refractive index of the cladding n_2. At their interface (core/cladding) the modes are kept in the core of the optical fiber by continuous total reflection.

The coating is understood to be the layer applied directly to the cladding surface during production. It must be removable for light coupling in and out of the fiber or for jointing fibers. It can be composed of several layers of plastic and must be applied evenly over the entire length without bare spots or variations in thickness. The coating may be colored and if necessary may have additional ring markings; it has a refractive index higher than that of the cladding, which means that undesirable light launched into the cladding is absorbed in the plastic within a few meters.

Examples of the Refractive Indices of an optical fiber

Core	1.48
Cladding	1.46
Coating	1.52

Mechanically, the coating should

protect the optical fiber from external influences and absorb shear forces which might bend the fiber in the microrange and thereby cause additional attenuation.

6.1 Multimode Fiber of Fused Silica Glass

The standard geometric values for multimode fibers are shown in Table 6.1 a and the categories in general use for multimode fibers in Table 6.1 b.

At a diameter of only 50 µm the core of a multimode fiber has the dimensions of a human hair.

Multimode fibers of category A1 with a core material of glass, a cladding material of glass and a profile index $g < 3$ (Chapter 4.2) are typically available with core/cladding diameters of 50/125 µm, 62.5/125 µm and 85/125 µm.

Table 6.1a Multimode fibers, dimensions

Fiber type G50/125	EN 188200, VDE 0888 Part 105, ITU-T Recommendation G. 651, IEC 60793-2
Core diameter	
Nominal value	50 µm
Permissible deviation	± 3 µm
Max. noncircularity	6%
Cladding diameter	
Nominal value	125 µm
Permissible deviation	± 3 µm
Max. noncircularity	2%
Max. concentricity error	6%

Table 6.1b Multimode fibers, categories according to IEC 60793-1-1

Category	Core/cladding material	Fiber type	Profile exponent
A 1	Glass/glass	Graded index	$1 \leq g < 3$
A 2.1	Glass/glass	Quasi step index	$3 \leq g < 10$
A 2.2	Glass/glass	Step index	$10 \leq g < \infty$
A 3	Glass/plastic	Step index	$10 \leq g < \infty$
A 4	Plastic/plastic	Plastic fiber	

Table 6.1c Multimode fibers for short distance links

Category	A2	A3	A4
Core/cladding diameter (µm)	100/140 200/240 200/280	200/300 200/380 200/230	980/1000 735/750 485/500
Attenuation	≤ 10 dB/km	≤ 10 dB/km	≤ 40 dB/100 m
Theoret. numerical aperture	0.23 to 0.26	0.4	0.5
Wavelength (nm)	850	850	650 to 660
Bandwith (MHz)	≤ 10 for 1 km	≤ 5 for 1 km	≤ 10 for 100 m
Typical distance	Up to 2 km	Up to 1 km	Up to 100 m

Figures 6.1 and 6.3 show the structure of a category A1 multimode fiber.

Table 6.1c shows the dimensional and transmission parameters of multimode fibers (categories A2, A3 and A4) for short links.

Codes (as used in Germany)

F – G 50/125 … … …

F	optical fiber
G	graded index fiber
50/125	nominal values: core diameter in µm/cladding diameter in µm

6.2 Single-Mode Fiber of Fused Silica Glass

In contrast to the multimode fiber, it is sufficient for a single-mode fiber (Figures 6.2 and 6.3) to state the wavelength-dependent mode field diameter $2w_0$ instead of the core diameter. At a wavelength of 1300 nm the mode field diameter is 10 to 12 per cent larger than the core diameter.

This dependence on the wavelength is particularly important in the fiber jointing. The quality of a splice – because only mode propagates in the optical fiber – is influenced more by the light-technical adjustment than by the glass-technical one.

The standard geometric values for single-mode fibers are shown in Table 6.2a, the categories in general use in Table 6.2b and some typical values in Table 6.2c.

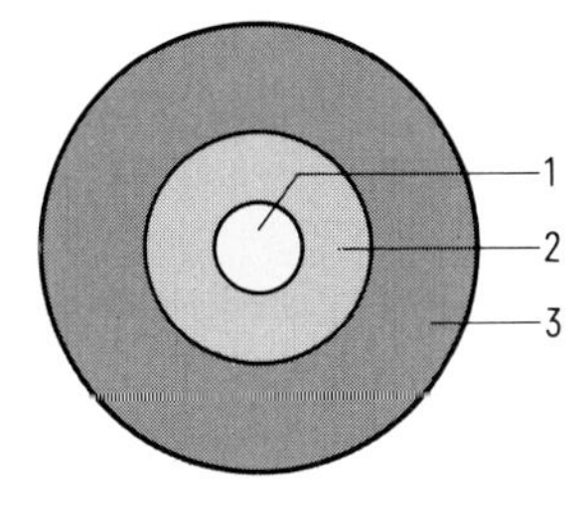

1 Core
2 Cladding
3 Coating

Figure 6.1 Multimode fiber

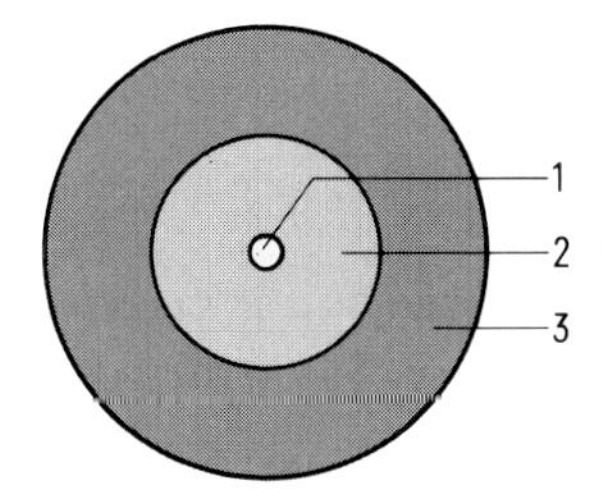

1 Core
2 Cladding
3 Coating

Figure 6.2 Single-mode fiber

 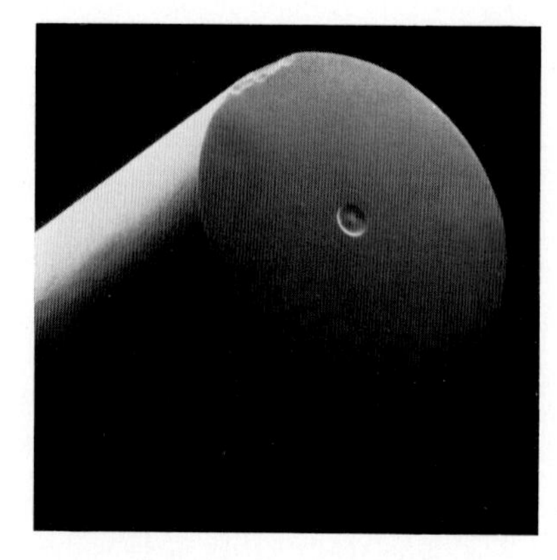

Figure 6.3 Etched end faces of a multimode and single-mode fiber

A mode field diameter of 10 µm at 1310 nm is commonly employed for single-mode fibers with matched cladding design (Figure 4.12a), and a value of 9 µm with depressed cladding design (Figure 4.12b).

The most widely used single-mode fiber is the category B1.1 fiber, as specified in ITU-T G.652. The B1.2 fiber is an optical fiber with a pure silica core, and specified in ITU-T G.654. The dispersion shifted fiber, category B2, specified in ITU-T G.653 is not recommended anymore for deployment. Category B3 single-mode fibers are characterized by a low dispersion over a large wavelength range. They are not used anymore in practical applications. Category B4, the nonzero dispersion shifted fiber (NZDSF), specified in ITU-

Table 6.2a Single-mode fibres, dimensiones

Single-mode fibers	EN 188100, VDE 0888 Part 102, ITU-T Recommendation G. 652, IEC 60793-2
Mode field diameter at 1300 nm	
Nominal value	8.6 µm to 9.5 µm
Permissible deviation	+/– 10%
Cladding diameter	
Nominal value	125 µm
Permissible deviation	+/– 2 µm
Max. noncircularity	2%
Max. concentricity error	1 µm[1])

[1]) Depending on the splice technique and the permissible splice loss values up to 3 µm may be selected.

Table 6.2b Single-mode fibers, categories according to IEC 60973-1-1

Category	Description	Zero dispersion wavelength Nominal value (nm)	Operating wavelength Nominal value (nm)
B 1.1	Dispersion unshifted	1310	1310
B 1.2	Loss minimized	1310	1550
B 2	Dispersion shifted	1550	1550
B 3	Dispersion flattened		1310 and 1550
B 4	Nonzero dispersion		1550

Table 6.2c Single-mode fibers, typical values

Category	B 1.1 (G.652)	B 2 (G.653)	B 4 (G.655)
Attenuation at 155 nm (dB/km)	0.20 to 0.22	0.22 to 0.24	0.20 to 0.23
Dispersion at 1550 nm (ps/nm · km)	17	0 to 2.7	2 to 3
Nonlinearity (Chapter 2.8)	good	bad	very good
PMD (Chapter 5.10)	good	good	good

T G.655, is the fiber of choice for long-haul high capacity (bit rates of more than 2.5 Gbit/s) transport systems employing Erbium-doped fiber amplifiers (EDFA) together with Time Division Multiplexing (TDM) and/or Dense Wavelength Division Multiplexing (DWDM) (Chapter 13).

Codes (as used in Germany)

F – E9/125 … … …

F optical fiber

E single-mode fiber

9/125 nominal values:
mode field diameter in µm/cladding diameter in µm

In addition to the construction already mentioned for single- and multimode fibers that have been standardized both nationally and internationally, there are other dimensions to meet special applications.

6.3 Technical Properties of Single-Mode and Multimode Fibers

Tensile Load

The maximum tensile load which can be carried for a duration of approximately 1 s is at least 5 N. It must either be tested in-line during production or off-line in a separate step. The entire length of fiber is subjected to a tensile proof stress larger than or equal to 100 kpsi = 0.7 GN/m^2.

Color Coding

In order to identify individual fibers, they are color coded and – if necessary – printed with ring markings and colors according to IEC 60304 and DIN 47002 (Table 6.3). The colors are applied in such a manner that they have no influence on the optical properties of the fibers.

Table 6.3 Color codes

Colors	
Blue Red Orange Black Green Yellow Brown Violet Gray Pink White Turquoise	according to IEC 60304 and DIN 47002

Temperature Range

Transportation and storage temperatures	–40 °C to 55 °C
Operating temperature	–60 °C to 85 °C

If for special applications these temperatures ranges may be exceeded, the manufacturer should be consulted.

Besides these values, IEC 60793-2 and VDE 0888 Part 101 stipulate a great number of mechanical, transmission and optical tests.

6.4 Optical Fiber of Plastic Material

Optical fibers made of transparent plastic materials are suitable for transmission of information over short distances of up to 100 m.

The most frequently used fiber type is composed of a polymethylmetacrylate (PMMA) core with a thickness of ca. 980 μm and a carbon polymer cladding with a thickness of approx. 20 μm (Figure 6.4).

Using the two refractive indices of 1.492 (core) and 1.417 (cladding), results in a numerical aperture of 0.47 and an acceptance angle of ± 28°. Therefore such a plastic optical fiber has very efficient light coupling.

Figure 6.5 shows the spectral attenuation of a plastic optical fiber.

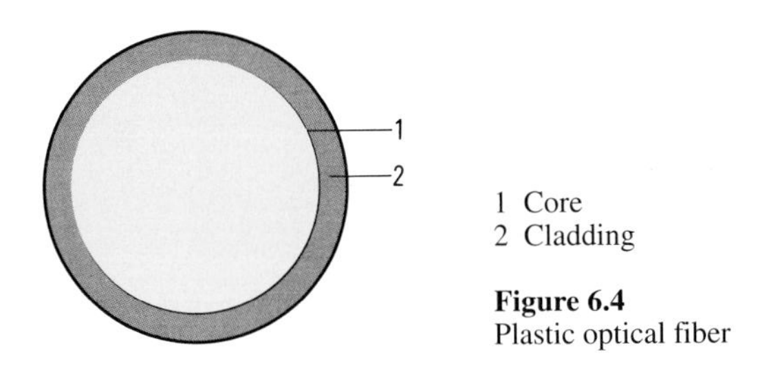

1 Core
2 Cladding

Figure 6.4
Plastic optical fiber

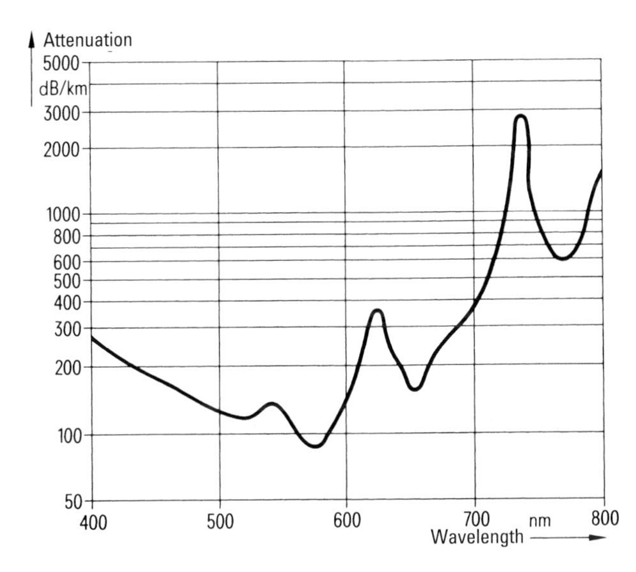

Figure 6.5
Spectral attenuation of a plastic optical fiber

Codes (as used in Germany)

F-P 980/1000 150:

F	Fiber
P	Plastic optical fiber with step index profile
980/1000	Core/cladding diameter in µm
150	Attenuation coefficient in dB/km at 650 nm

Temperature Range

Transportation and storage temperature	– 20 °C to 50 °C
Operating temperature	0 °C to 55 °C

7 Manufacturing of Optical Fibers

Optical fibers are usually manufactured in several production steps, making it possible to selectively optimize mechanical, geometric and optical properties of the fiber. This procedure also allows fast and economic mass production, one of the important preconditions of modern optical communication cables technology.

In all of the techniques in use today a preform is produced first. This is a glass rod consisting of core and cladding glass. The cross-section of the preform shows a scale model of the geometric dimensions and refractive index profile of the optical fiber which can be made from it. While one end is heated at a higher temperature the preform is drawn out to a fiber and at the same time the coating is applied to the fiber as a protective cover.

7.1 Preform Production by Melting Glass

One of the first techniques used in the production of optical fibers is the *rod-in-tube* method, whereby an ultrapure fused silica glass rod as the core is slid into a fused silica glass tube with a lower refractive index as the cladding. The dimensions of rod and tube are such that little gap remains between them. This technique has the disadvantage that minute perturbations and impurities at the interface cause very high attenuation (500 to 1000 dB/km) when the fiber has been drawn and that only multimode step index fibers can be produced in this manner.

In order to avoid problems, a different method is used in which the core and cladding glass are brought together in a molten state and the fibers are drawn directly out of the molten state. This method is called the *double crucible* or compound melting method because the glass for the core and the cladding are molten in two separate crucibles. Types of glass used are ultrapure multi-component glasses, e.g. lead alkali silicate glass and sodium borosilicate glass. By diffusion or ion exchange between the core and cladding glass it is also possible to produce optical fibers with graded index profiles (the Selfoc method). Because the insides of the crucibles cannot be kept absolutely clean, in addition to transition metals impurities enter the fiber and increase the attenuation (5 to 20 dB/km at 850 nm). This method is used in particular for fat core fibers (core diameter ≥ 200 μm).

In the *phase separable glass* or *phasil method* a sodium borosilicate glass rod is formed at 1200 °C and held at 600 °C for several hours, during which time a separation of a sodium borate glass phase in a SiO_2 glass matrix takes place. Transition metals such as Fe and Cu collect in the sodium borate glass phase and can be leached out with an acid so that a porous preform remains. This is soaked in an ultrapure salt solution, e.g. cesium nitrate, and afterwards the outside is washed. The cesium doping increases the refractive index inside. The washed region becomes the cladding. Optical fibers with step index and graded index profiles with an attenuation of 10 to 50 dB/km at 850 nm can be produced by this method.

A further method of preform production involves the use of a rod of commercially available fused silica glass as the core and during drawing a cladding of transparent plastic (plastic clad) with a low refractive index is applied. Attenuation values ranging from 5 to 50 dB/km at 850 nm have been achieved. There are optical fibers made purely from plastic materials, the so-called *plastic fibers*, whereby core and cladding are produced from optically pure plastics with different refractive indices. Attenuation values of 100 to 400 dB/km around 600 nm have been measured with them (Chapter 6.4).

7.2 Preform Production by Glass Deposition out of the Vapor Phase

The breakthrough in the production of optical fibers with extremely low attenuation was only achieved with the different *vapor phase deposition methods*, a technique which was first used in 1970 by Corning Inc., USA.

The glass deposition can take place on the outside surface of a rotating target rod (OVD method, *o*utside *v*apor *d*eposition), on the end face of a fused silica glass rod (VAD method, *v*apor *a*xial *d*eposition) or on the inside surface of a rotating fused silica glass tube (IVD method, *i*nside *v*apor *d*eposition). For the inside vapor deposition method the necessary energy for the glass deposition can be gained either from outside by an oxyhydrogen gas burner (MCVD method, *m*odified *c*hemical *v*apor *d*eposition) or from inside with a plasma flame (PCVD method, *p*lasma-activated *c*hemical *v*apor *d*eposition).

In these techniques (Figure 7.1) glass deposition occurs by the reaction of volatile, ultrapure compounds in an oxyhydrogen or plasma flame, as described in Chapter 3.1.1. These methods are used today in modern industrial mass production to manufacture preforms for multimode and single-mode fibers with extremely low attenuation (0.18 dB/km at 1550 nm) and high bandwidth (>2 GHz km at 1300 nm), or low dispersion (<3.5 ps/(nm km)) between 1285 and 1330 nm. On the one hand, particular emphasis is placed on improving the deposition rate, i.e. the amount of the deposited particles (soot particles) per minute, which typically lies in the range of 1 to 5 g/min. On the other hand, it is desirable to increase the size of the preform in order to be able to draw fibers with a length of more than 100 km.

7.2.1 OVD Method

Preform production by the OVD method takes place in two steps. First a *target rod* of fused silica glass, Al_2O_3 or graphite is rotated about its longitudinal axis in a turning lathe and heated on the *outside* in a narrow zone by an oxyhydrogen or propane gas burner (Figure 7.2).

Oxygen (O_2) and whatever dopants are required for the refractive index profile, e.g. the metallic halides ($SiCl_4$, $GeCl_4$, BCl_3, PCl_3), are fed into the burner and transformed into the corresponding oxides. These form a fine soot which is deposited on the rotating rod. If the rod is also moved back and forth lengthwise, then a *porous glass preform* is built up in layers. Each layer can be doped differently, i.e. a certain amount of impurity material can be added to the basic material, SiO_2. For graded index profiles the doping of the core with GeO_2 is reduced from the first layer continuously until pure SiO_2 is deposited for the cladding. In the case of step index profiles the doping of the core with GeO_2 remains constant layer after layer.

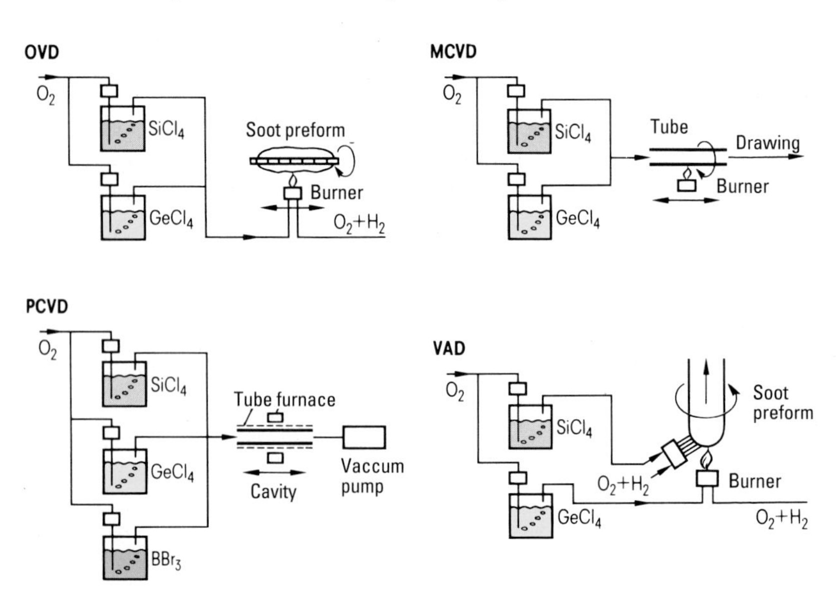

Figure 7.1 Preform production by glass deposition out of the vapor phase

As soon as enough layers have been deposited for the core and cladding of the fiber, the process is stopped and the cylindrical tube made of soot particles is removed from the target rod. In the second step of this method the tube preform is heated in segments over its entire length (Figure 7.2). At temperatures between 1400 and 1600 °C it collapses (shrinks) to a solid, bubble-free, transparent glass rod (blank), in which the center hole is closed. During sintering the preform is constantly flooded with gaseous chlorine as a drying agent in order to remove all possible traces of water from the glass, because otherwise higher attenuation would have to be expected.

7.2.2 VAD Method

In the VAD method the soot particles created in an oxyhydrogen burner are deposited on the *end face* of a rotating fused silica glass rod (Figure 7.3).

The porous preform created in this way is drawn upward, whereby the distance between the burner and the preform which is growing axially must remain constant. Several burners can be used simultaneously in order to create the refractive index for the core and cladding. Depending on the construction of the burners and their distance from the rod and the temperature during deposition, different refractive index profiles can be produced. Fol-

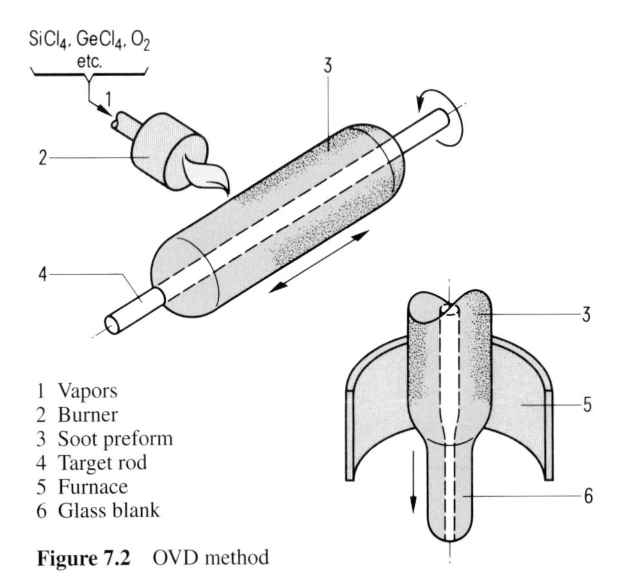

Figure 7.2 OVD method

lowing the deposition the porous preform is collapsed (consolidated) to a transparent preform (blank) by a ring-shaped furnace. Gaseous chlorine is flooded around the collapsing preform to dry it.

In order to obtain fibers with a thicker cladding, it is possible then to slide a fused silica glass tube as a cladding over the preform similar to the rod-in-tube method.

7.2.3 MCVD Method

Preforms are produced in two steps with the MCVD method. First a fused silica glass tube is rotated around its longitudinal axis in a lathe and its outside is heated in a narrow zone by an oxyhydrogen gas burner moving along the outside of the tube (Figure 7.4).

Oxygen and the gaseous halide compounds ($SiCl_4$, $GeCl_4$, PCl_3) needed as dopants flow through the tube. These halides do not react in the flame of the burner as in the OVD and VAD methods, but rather *inside* the tube. This leads to a deposition of many thin layers of glass on the inside of the tube which can be doped to create the desired refractive index profile. The tube

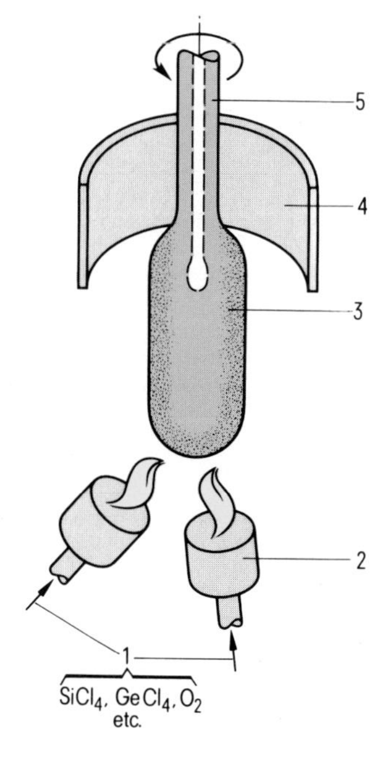

1 Vapors
2 Burner
3 Soot preform
4 Furnace
5 Transparent preform

Figure 7.3 VAD method

itself is the outer part of the cladding glass. Through vapor phase deposition the inner part of the cladding glass and the core glass are produced.

A layer of glass is created when fine particles develop at 1600 °C in the heated zone and are deposited on the inner wall of the tube. When the burner is moved further in the direction of flow, the deposited particles are fused into a thin layer of glass.

In the second step of the MCVD method the tube, onto which the required layers have been deposited, is heated to 2000 °C in segments along its length. Thereby the tube shrinks to a rod. If the gases reacting inside the tube have been kept free of hydrogen, then no special drying procedures are necessary for this method because the gases used for heating, which are usually high in hydrogen, only come in contact with the outside of the tube and any other environmental influences also have no effect.

7.2.4 PCVD Method

The production of preforms by the PCVD method follows basically the same process steps as described for the MCVD method (Chapter 7.2.3). The difference lies in the reaction technique. A plasma is created when gas is

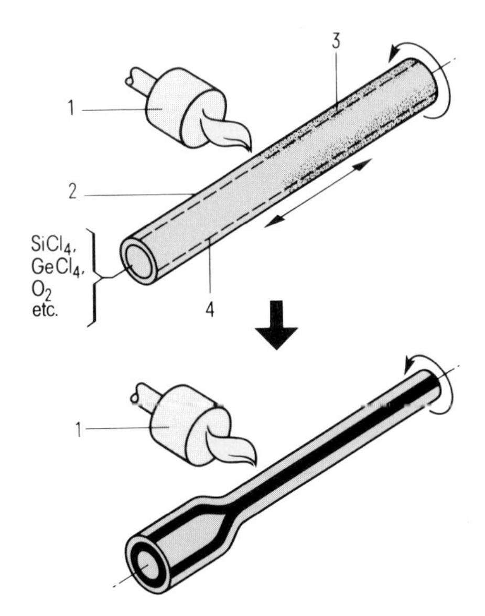

1 Burner
2 Glass tube
3 Soot layer
4 Glass layer

Figure 7.4 MCVD method

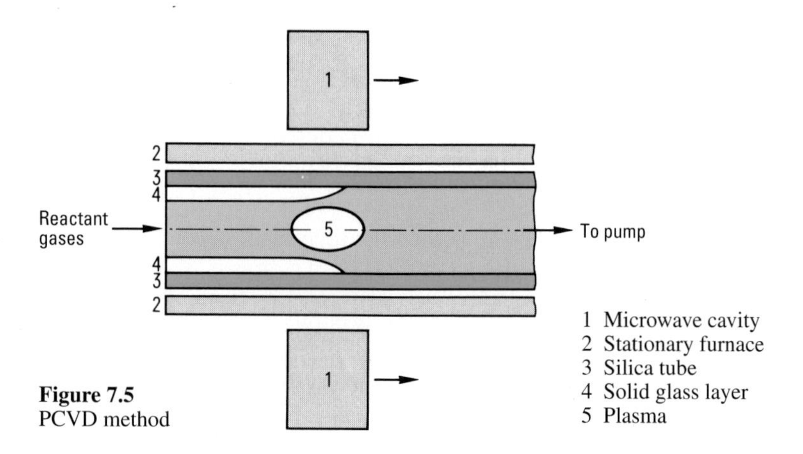

Figure 7.5
PCVD method

excited, for example, by microwaves. This ionizes the gas, i.e. it is separated into its electric charge carriers. When these charge carriers are reunited thermal energy is freed which can be used to melt materials with a high melting point. With the plasma method the halides react with the help of a low-pressure plasma (pressure in the gas is approximately 10 mbar) and with oxygen it builds SiO_2. The soot particles produced in such a way are deposited directly at about 1000 °C as a layer of glass (Figure 7.5).

Because the plasma flame moves quickly up and down the tube, more than a thousand very thin layers can be produced, which allows an increase in the precision of the refractive index profile.

7.3 Fiber Drawing

In order to draw out a fiber, the preform is attached to a mount in a draw tower (Figure 7.6).

The mount is vertically adjustable with a feed mechanism. The lower end of the preform is heated to 2000 °C by a heating element, such that the fiber can be drawn down from the melting preform. In order for the diameter of the fiber to remain constant at the required value, both drawing speed (typically 300 m/min) and the feed mechanism must be precisely adjustable through an automatic control system.

During drawing the geometric proportions of core and cladding glass are retained even though a reduction in diameter from the preform to the fiber of about 300 to 1 is possible. The refractive index profile also remains unchanged.

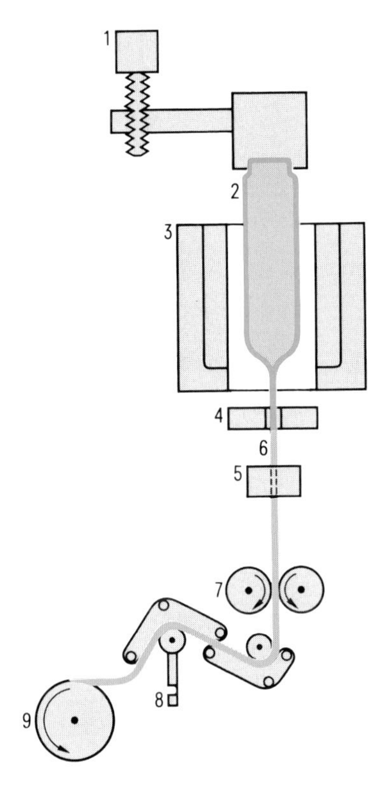

1 Downfeed mechanism
2 Preform
3 Furnace
4 Diameter monitor
5 Coater
6 Waveguide
7 Drawing tractors
8 Tensile strength monitor
9 Winding drum

Figure 7.6
Schematic of fiber drawing

Directly after diameter monitor, the coating is applied to the fiber. This plastic coating, which usually has a two layer structure, serves to improve the fiber strength, whereby the inner layer protects against microbending and the outer layer simplifies handling. When the coating has been hardened by heating or UV curing, the tensile strength of the fiber is checked as it passes through an in-line test monitor. The coated fiber runs over an array of rollers which apply a precisely measured tensile force to it. The fiber must withstand this minimum stress before it is wound onto a cylindrical drum.

8 Optical Fiber Buffers

In most applications single-mode and multimode fibers as such cannot be used without further enhancement. This is due in a large part to their relatively low breaking elongation of a few tenths of a percent and to the fact that very high increases in attenuation result from tensile, bending and torsion strain. Only when the fiber is protected by a buffer can it be put to practical use.

This goal, i.e. to protect the optical fiber as much as possible from external influences and to keep it functioning within mechanically permissible limits, is achieved through design measures in cable technology.

On the one hand there are special constructions for protecting the fiber, the buffer or the slot; on the other hand there are methods (described below) to incorporate tensile and antibuckling members in the cable, as well as the correct choice of the stranding, all of which make it possible to tailor the entire structure of the cable to the application.

8.1 Single-Fiber Loose Buffer

A single-fiber loose buffer is a small plastic tube with only one optical fiber in it, which is sufficiently protected against deformation and friction. The buffer must hold its shape, be tough, not be susceptible to aging and be very flexible, so that it can be handled in much the same way as quads or coaxial pairs in traditional copper cables, without noticeably stressing the optical fiber. The single-fiber loose buffer therefore has all the characteristics of a fundamental construction element which can be used universally.

The buffer tube is composed of an inner protective layer with a very low friction coefficient and an outer layer to shield the fiber against mechanical influences. By means of different basic materials or combinations of materials (e.g. polyester and polyamide) it is possible to compensate for a wide range of environmental conditions.

The optical fiber itself with some tenths of a millimeter of free space and a defined length lies in this protective buffer tube, which also permits a radial mobility of the fiber. The buffer tube is smooth inside, causing the lowest possible resistance to movement of the fiber.

Assuming the optical fiber and the buffer tube to have the same length, then with appropriate stranding the single-fiber loose buffer can be seen as an excess length which can easily permit 0.4 per cent of change in the length of the cable and can be increased by a maximum of twice that without buckling or stretching the fiber.

A special advantage of the loose buffer is that it can easily be stripped for joints or light launching. This is very useful when preparations are made for splices or connectors.

Figures 8.1 to 8.3 clarify the advantages of the single-fiber loose buffer by showing the position of the optical fiber under different conditions.

Figure 8.1 shows the position of the completely unstressed optical fiber in the loose buffer. The lengths of the fiber and the buffer tube are equal.

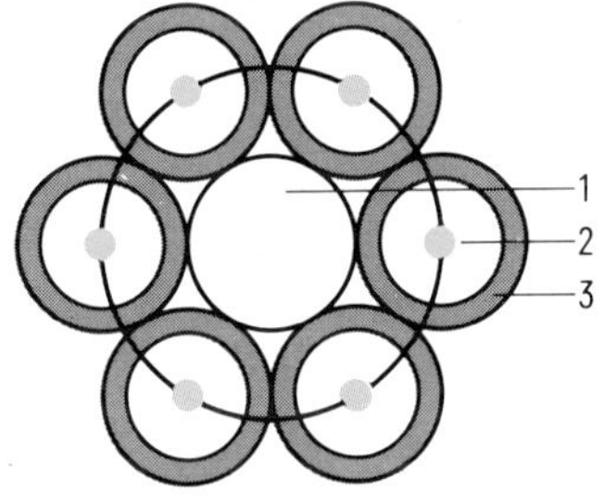

1 Central member
2 Optical fiber
3 Buffer tube

◁ **Figure 8.1**
Position of the unstressed optical fiber in the loose buffer

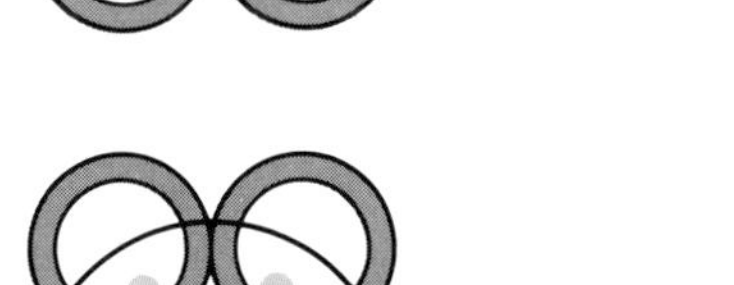

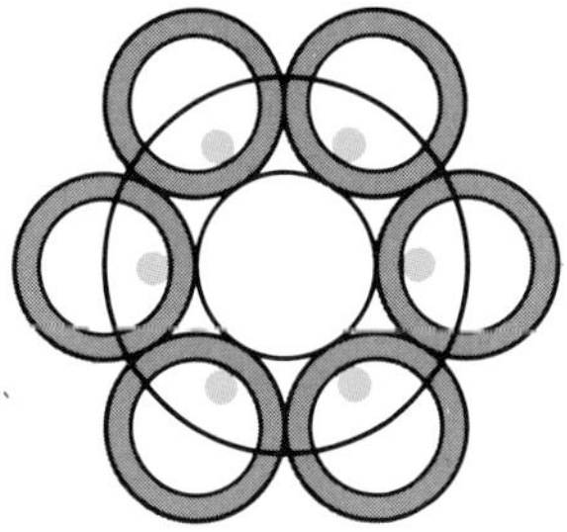

Figure 8.2
Position of the optical fiber when the cable is elongated

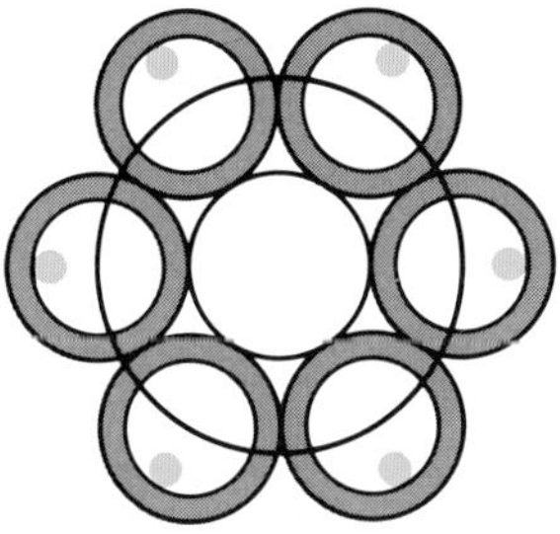

Figure 8.3
Position of the optical fiber when the cable is contracted

When the optical fiber cable is elongated, e.g. due to tensile stress, the fiber moves toward the inside of the buffer tube, at first without touching it or being deformed – which would cause higher attenuation. Depending on the design of the loose buffer the elongation of the cable will only affect the optical fiber when the strain reaches approximately 0.5 to 1 per cent. Only values higher than these will result in attenuation increases in the fiber (Figure 8.2).

If a cable is contracted, the optical fiber moves to the outside of the buffer tube (Figure 8.3). Here again the excess length makes it possible to adjust to external conditions without resultant higher attenuation. For example, cables contract due to cooling in winter.

Filling Compound

When an optical cable is damaged, it must be expected that, under certain environmental conditions, water will enter the buffer tube and may run through it. Because the water could freeze and depending on the amount, expand differently, causing local volume variations and thereby applying pressure to the optical fiber in many places, there is a risk of *microbending* and undesirable increases in attenuation. To prevent this the buffer tube is filled with a compound (Figure 8.4).

This compound is slightly thixotropic[1)] and chemically neutral, will neither freeze nor drip out at the temperature range of – 30 °C to + 70 °C which might prevail and also will not corrode the coating of the optical fiber or cause it to swell. The compound is easy to wipe and wash off and leaves no residue that would make it difficult to connect the fibers; it also contains no highly inflammable materials.

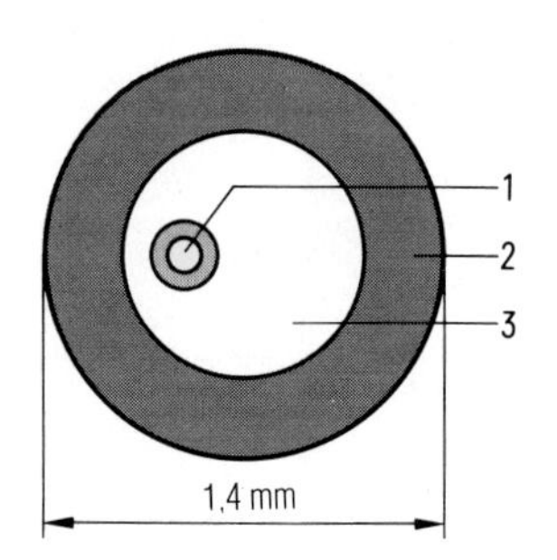

1 Optical fiber
2 Buffer tube
3 Filling compound

Figure 8.4
Single-fiber loose buffer

[1)] Thixotropic: a property of some gels which liquefy when they are stirred and solidify when they are at rest again.

8.1.1 Production

In the manufacturing process two extruders are aligned back to back, which permits production of the loose buffer's inner and outer tube in one continuous process. For this purpose a control system is needed that functions very precisely and guarantees a constant extrusion rate of the buffer tube materials at about 250 °C, if the necessary wall thickness of the tubes of only a few tenths of a millimeter is to be maintained (Figure 8.5).

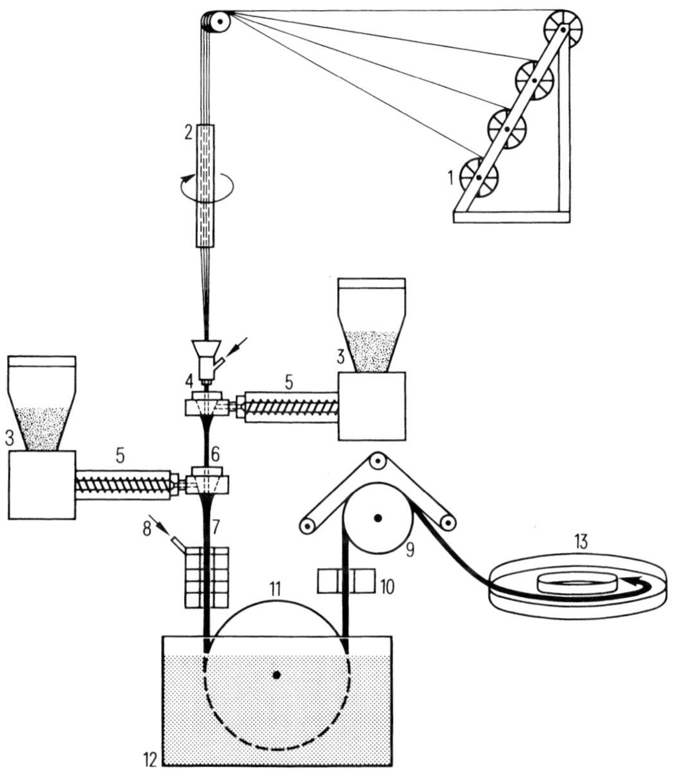

1 Spools of optical fiber
2 Stranding of fibers
3 Plastic in pellet form
4 Injection of filling compound
5 Screw (extruder)
6 First buffer tube
7 Second buffer tube
8 Cooling water
9 Capstan
10 Diameter monitor
11 Sheave
12 Cooling basin
13 Take-up tray

Figure 8.5 Schematic of the buffer tube production

The advantages of this two-layer buffer tube over a single-layer buffer tube lie in the greater degree of freedom in the choice of materials and their combinations. Thermal and mechanical problems are overcome more easily through these possibilities of variations.

Because the filling compound must be free of any impurity, it is applied under constant pressure directly into the tube through an injection needle during the production process of the buffer tube.

The most important aspect of the production process for loose buffers is the precise matching of the length of the buffer tube and the optical fiber.

Traditional spools onto which copper wires are wound are not suitable for use with loose buffers. This is due partly to their limited capacity and partly to the fact that, during take-up, layers can slip together. This would lead to uncontrolled pressure conditions among the loose buffers and thereby to damage to them. Furthermore, this kind of storage would make it difficult to match the lengths of the optical fibers precisely to that of the buffer tubes. For this reason loose buffers are wound onto large horizontal trays located directly next to the production machines. These have a capacity for taking up high volumes (up to several kilometers).

8.2 Multifiber Loose Buffer

As described in Chapter 8.1, loose buffers have already proved useful for practical applications with only one optical fiber. These tubes, which come with an outside diameter of 1.4 mm, are used in cable constructions with low fiber counts.

With the single-fiber loose buffers it is possible to design and produce cables even with a greater number of optical fibers. However, the cable constructions become increasingly complex, their outside diameters become comparatively large and correspondingly their weights increase greatly. All this makes their practical application ever more difficult and inconvenient.

In order to limit these disadvantages, instead of putting one optical fiber in a buffer tube, between two and twelve single-mode or multimode fibers are inserted into a somewhat larger buffer tube (outside diameter 1.8 to 3.5 mm), the so-called multifiber loose buffer.

As with the single-fiber loose buffer, the hollow space is filled with a flexible, slightly thixotropic compound that neither drips nor freezes in the temperature range from – 30 °C to + 70 °C (Figure 8.6).

The *maxitube* is a further development of the multifiber loose buffer. It can hold a large number of optical fibers either individually or in form of fiber bundles or fiber ribbons. The optical fibers inside this larger buffer tube (typical outside diameter 6 mm) are stranded. Through the stranding of the fibers *inside* the buffer tube an excess length of approximately 0.5% is created (Figure 8.7). This multifiber maxitube is used, for example, as a basic element for self-supporting aerial optical cables.

8.2.1 Production

The multifiber loose buffer is produced similarly to the single-fiber loose buffer (Figure 8.5). The only difference lies in the dimension of the buffer tube. As an example, for single-fiber loose buffers according to VDE 0888, Part 3 the ratio of outside diameter to inside diameter is 1.4 mm/0.85 mm and for multifiber loose buffers it is 3.0 mm/1.8 mm, also as an example.

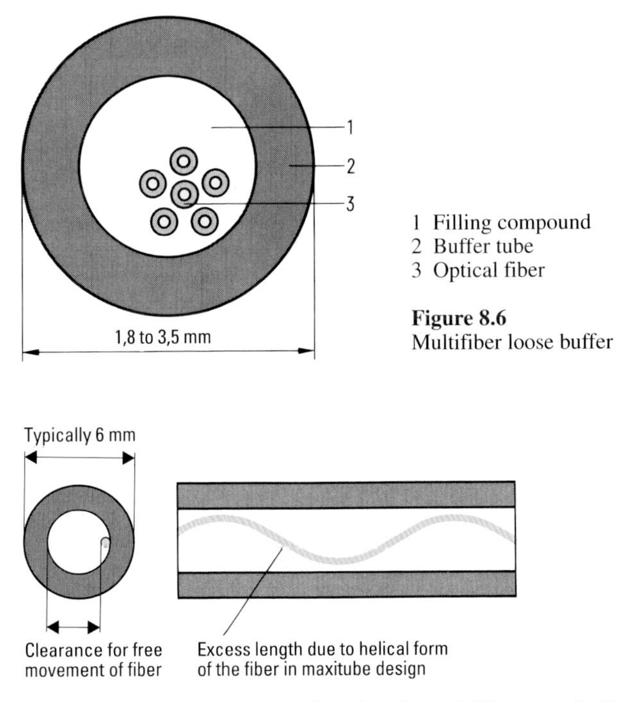

1 Filling compound
2 Buffer tube
3 Optical fiber

Figure 8.6
Multifiber loose buffer

Figure 8.7 Optical fiber excess length in the multifiber loose buffer in maxitube design

8.2.2 Technical Characteristics of Single-Fiber and Multifiber Loose Buffers

In the following, examples for the technical characteristics of single-fiber and multifiber loose buffers are presented: Dimensions and color codes.

Dimensions

The guiding values for the outside diameter of loose buffers are given in Table 8.1.

The wall thickness of the loose buffers is approximately 20 per cent of the outside diameter of the loose buffers.

In addition to these standardized values others could be used for practice depending on the application, e.g. for longer installation lengths or for self-supporting aerial cables with a layer structure.

Color Codes

Single-fiber and multifiber loose buffers are usually colored. If one or several layers of loose buffers are stranded around a central member, in each layer one of the loose buffers can be, for example, red. With the aid of this red buffer tube, the so-called marking element, it is possible to identify all other loose buffers of that layer. While facing the cable end (A end) which is lying on top of the cable drum, one counts clockwise beginning with the red buffer tube. The next buffer tube is green and all other tubes are natural colored.

In Germany, for optical fibers and for loose buffers, the National Standards Institute can issue its approval logo which can only be used after an appropriate application has been filed and tested according to their procedures.

Table 8.1
Typical values for the diameter of loose buffers

	No. of fibers	Outside diameters (mm)
Single-fiber loose buffer	1	1.4
Multifiber loose buffer	2 - 12	1.8 - 3.5
Multifiber maxitube	>12	6

8.2.3 Applications of Single-Fiber and Multifiber Loose Buffers

Optical cables with single-fiber and multifiber buffers are used especially in situations where high-quality transmission is expected in spite of various environmental influences that may make it difficult to meet these expectations.

Outdoor cables with single- or multimode fibers are therefore mainly designed with single-fiber or multifiber loose buffers. Temperature variations between – 30 °C and 70 °C and the usual mechanical influence such as stress, pressure and buckling forces can be handled with cable types of standard design.

A further advantage of this type of design is that when demands exceeding the values normally expected are placed on the cable, e.g. extreme environmental conditions, other dimensions and/or a more suitable material or combinations of materials can be chosen for the buffer tube to meet these special applications. These could be, for example, self-supporting aerial cables, submarine cables or shaft cables in mining operations.

These cables with loose buffers have been accepted by the Deutsche Telekom AG and other operators of telecommunication networks within Germany as well as by others abroad with a wide variety of application profiles.

8.3 Tight Buffered Fiber

A simple way to protect the optical fiber from external influences is to apply a tight buffer of suitable plastic directly over the coating of the fiber (Figure 8.8). This kind of buffer construction allows a considerable reduction of the outside diameter compared to the single-fiber loose buffer.

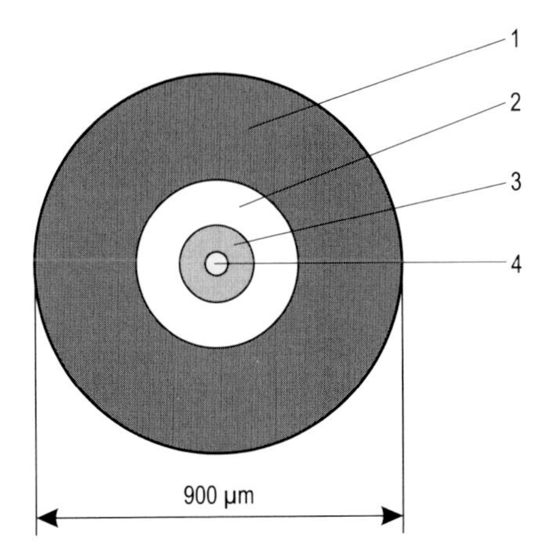

1 Plastic buffer
2 Fiber coating, 250 µm
3 Cladding
4 Core

Figure 8.8 Tight buffered fiber

However, in cable constructions with these tight buffered fibers elastic elongation of the cable due to high tensile stress directly affects the fiber because the necessary excess length is lacking, which is provided for in the loose buffer. This disadvantage can partly be compensated for by putting strength elements of aramid yarn into the cable.

On the other hand water penetration into the buffer can have negative effects on the fiber if the temperature falls below the freezing point. In that case, the freezing water exerts mechanical stress on the fiber, which can result in worse transmission properties or even in the destruction of the fiber. A loose buffered fiber has some excess length and is protected against water penetration due to the filling compound inside the buffer tube.

Compared to the loose buffered fibers, tight buffered fibers are of smaller outside diameter and lower weight. They are more flexible and much easier to strip, to install and to connectorize.

Especially the TBII tight buffered fiber sets new industry standards for user-friendliness as well as easy fiber installation and termination. It enables users to easily remove longer lengths of buffer – up to 20 cm – in a single pass. This is due to a thin layer of teflon between the 250 µm coating and the buffer. The TBII fiber attenuation values meet all known industry standards for optical and mechanical performance, such as EN 187000, EN 50173 and IEC 60794.

It is important to distinguish tight buffered fiber from the so-called "composite buffered fiber". The composite buffered fiber swims in a gliding layer with a radial clearance of 50 to 100 µm. This gliding layer is difficult to handle and to clean. Additionally it is highly flammable.

With the new generation of TBII tight buffered fiber, there is no more need for the composite buffered fiber for standard inhouse cabling.

8.3.1 Production

For tight buffer production the buffer is extruded directly onto the optical fiber which usually measures 250 µm in diameter. The buffered fiber has an outside diameter of 900 µm.

Similar to the production of loose buffers, there are horizontal and vertical methods of manufacturing this single layer of buffer on a fiber.

In order to enable easy stripping of the buffered fiber, a thin layer (5 to 10 µm) of teflon is sprayed onto the bare 250 µm fiber before the plastic buffer is extruded onto it.

For the purpose of identification, the buffer material can be color-coded.

Due to the direct extrusion of the buffer onto the fiber and the resulting minor stress on the fiber, the transmission performance is not quite the same as for loose buffered fiber. However, as the standard application length of a tight buffered fiber is not more than a few hundred meters, this deterioration of optical performance does not affect the applications at all.

8.3.2 Technical Characteristics

In the following, as an example, the technical characteristics of tight buffered fiber based on international (ISO/IEC 11801) and European (EN 50173) standards are given. Their transmission properties are listed in Table 8.2.

Dimensions

The nominal value for the diameter is 0.9 mm for single-mode and multimode tight buffered fibers and the permissible tolerance is ± 0.1 mm. The nominal weight is approx. 0.8 kg/km.

Mechanical Properties, Operating Temperature

The short-term tensile strength of a tight buffered fiber must be at least 5 N, while the long-term tensile strength should be at least 2.5 N. The minimum bend radius during installation (loaded) should not become less than 5 cm, while the bend radius after installation should be at least 3 cm. Operating temperature range is between – 5° C up to + 70 °C, with a maximum change in attenuation of 1 dB/km in that temperature range.

The maximum permissible tensile stress is 400 N (depending on fiber count).

Table 8.2
Transmission properties of tight buffered fibers

Fiber type	Wavelength (nm)	Attenuation (dB/km)	Bandwidth (MHz × km)
Multimode 62.5 μm	850	3.5	200
	1300	1.0	500
Single-mode	1310	< 1.0	–
	1550	< 1.0	–

8.3.3 Applications

The tight buffered fiber can be used in all cases where short connections of a line or jumper type are called for; e.g. for pigtails for internal cabling in cabinets or shelves. It also can be deployed as transmission media inside machines, computers etc., where different fiber optic devices must be interconnected.

The tight buffered fiber should be used in a dry environment. Wet environment can only be accepted if there is a guarantee that the temperature does not drop below the freezing point.

8.3.4 Buffered Plastic Optical Fiber

In addition to unbuffered plastic optical fibers (Chapter 6.4), there are plastic optical fibers which have a protective plastic buffer to protect them from ex-

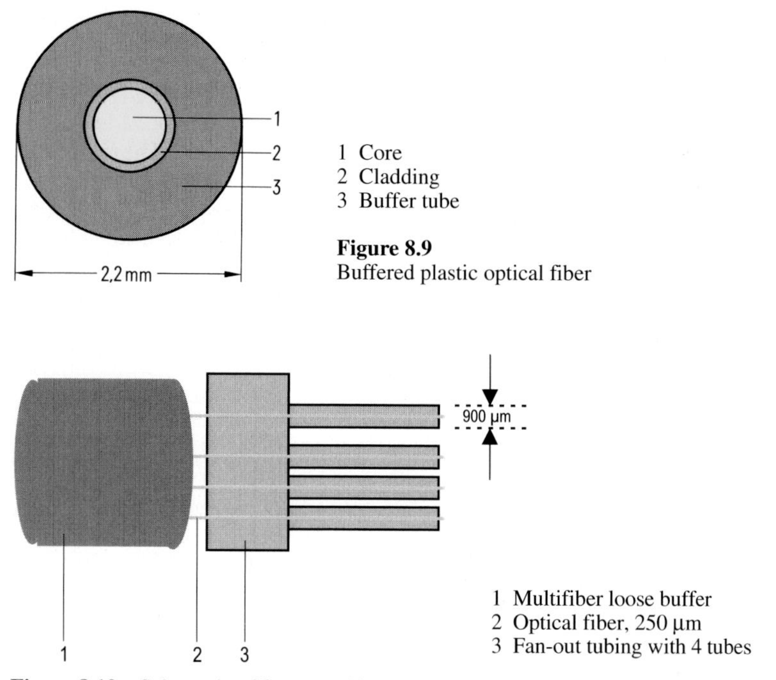

1 Core
2 Cladding
3 Buffer tube

Figure 8.9
Buffered plastic optical fiber

1 Multifiber loose buffer
2 Optical fiber, 250 µm
3 Fan-out tubing with 4 tubes

Figure 8.10 Schematic of fan-out tubing

ternal influences (Figure 8.9). The typical outside diameter of the buffer is 2.2 mm for a plastic optical fiber with a 1000 μm cladding diameter.

8.4 Fan-Out Tubing

A main advantage of a tight buffered fiber compared to a loose buffered fiber is the possibility to connectorize the fiber directly, i.e. a field installable connector can directly be terminated at the 900 μm tight buffered fiber. In order to provide direct termination to loose buffered fibers, a so-called *fan-out tube* (Figure 8.10) is used. A 250 μm loose buffered fiber will be inserted into the tube, which has an outer diameter of 900 μm and therefore is suited for direct termination with field installable connectors (without pigtail splicing).

8.5 Ribbon Technology

As described in Chapter 8.2, the multifiber loose buffer or maxitube has proved to be successful for enclosing several individual optical fibers. With respect to packing density and fiber jointing techniques, the ribbon technology can further improve this buffer design.

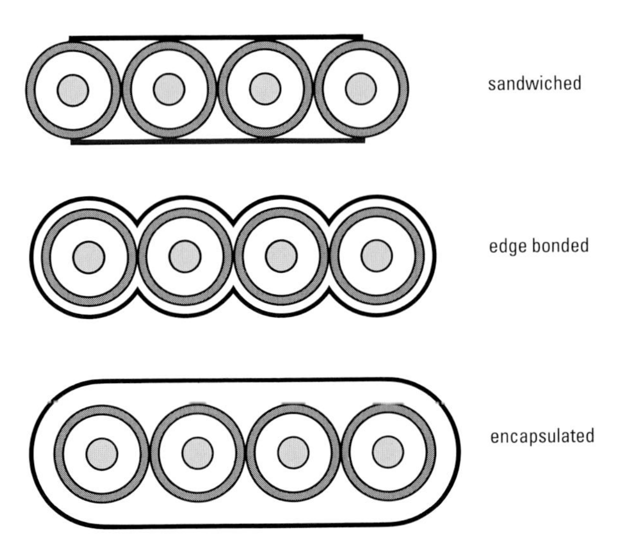

Figure 8.11 Optical fiber ribbon types

In this technology, 2 or more optical fibers are joined in an ordered unit to form a ribbon. The individual optical fibers are placed equally spaced and parallel to each other in one layer. Three ribbon types may be distinguished based on the way they are affixed: sandwiched, edge bonded and encapsulated (Figure 8.11).

In these methods the individual optical fibers are either bonded together in one layer between two polyester adhesive foils, or without foils they are glued together at their edges, or they are completely encapsulated in a plastic coating and thereby firmly affixed to one another. The ribbons can then be stacked on top of each other forming a stack with a rectangular cross-section. They can then be placed, e.g. in properly dimensioned slots of a slotted core cable or in maxitube (central loose tube) cables (Chapter 9.1.1).

9 Optical Cable Design

The highly different applications of optical fibers in communication cable technology require a variety of cable construction designs. The design of these cables specificies the corresponding dimensions and materials. Based on the structure of the fiber buffer, as described in Chapter 8, the cable core, the sheath and, as necessary, the armoring and protective cover are selected so that the optical cable will perform reliably over a long service life. Particular importance is placed on protecting the optical fibers in these cables against environmental influences such as temperature changes and mechanical stresses.

9.1 Cable Core

In order to increase the mechanical strength of optical cables with single-fiber and multifiber loose buffers, these buffers are stranded around a *central member*, whereby the central member serves both as a support (buckling protection, against kinking) and as a strain relief member. Due mainly to the *stranding*, the optical fibers have a well-defined free space within which strain, buckling, pressure, and of course bending stresses will have no influence on the transmission characteristics. In addition to single-fiber or multifiber loose buffers, tight buffered fibers or ribbon designs, *fillers* can also be stranded with them, i.e. buffer tubes without optical fibers or simply solid PE fillers, and also copper wires as pairs or quads. The entire configuration of the *stranding elements* in a cable, together with the antibuckling, the strength members and the wrapping around them, if used, is called the *cable core*.

9.1.1 Stranding

In optical cable technology cables are usually stranded in layers. Thereby the stranding elements are arranged concentrically in one or several layers around a central member (Figures 9.1 and 9.2). If the stranding elements are individual elements, e.g. single-fiber loose buffers, multifiber loose buffers, copper wires of fillers, then the cable is described as an optical *layer cable*; if on the other hand the cable core is composed of stranding elements which consist of units of stranded elements, then it is described as an optical *unit cable*.

In the trunk network of the telecommunication administrations, layer cables with multifiber loose buffers predominate (Figure 9.3). In local networks both cables stranded in layers (Figures 9.1 and 9.2) and unit cable constructions (Figure 9.4) are employed, because with a unit cable it is possible to greatly increase the packing density.

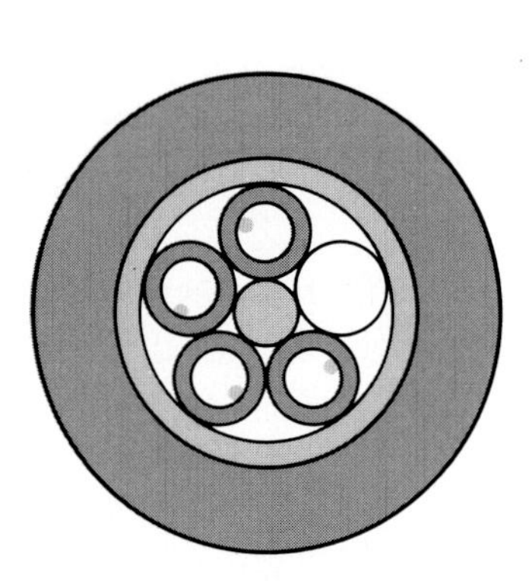

Figure 9.1
Layer cable with single fiber loose buffers in one layer

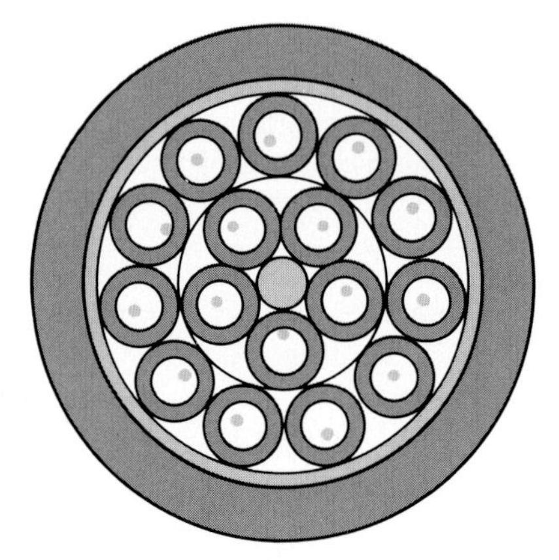

Figure 9.2
Layer cable with single fiber loose buffers in two layers

Figure 9.3
Layer cable with mulitfiber loose buffers in one layer

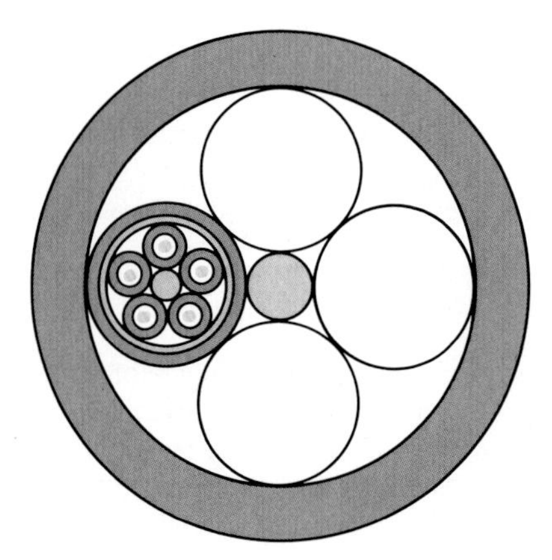

Figure 9.4
Unit cable

The geometrical relations in an optical cable with one layer of n stranding elements, all of which have a diameter D (Figure 9.5), can be described with the aid of the parameters a, b, c, e and the stranding angle α (cf. below) as follows:

$$a = \frac{1}{\sin\frac{\pi}{n}};$$

$$b = \frac{1}{\tan\frac{\pi}{n}} \qquad \text{with } b = 0, \text{ if } n = 2;$$

$$c = \frac{1}{\sin\alpha},$$

$$e = \sqrt{c^2 + \frac{1-c^2}{a^2}}$$

The *total outer diameter* of the cross-section is

$$D_1 = D \cdot (a \cdot e + 1)$$

and the *core diameter* is

$$d = D \cdot (a \cdot e - 1).$$

In addition the *outside interstice diameter* is

$$d_1 = \frac{D}{2} \cdot \frac{(1 + a - b)^2}{2 + a - b}$$

and the *inside interstice diameter* is

$$d_2 = \frac{D}{2} \cdot \frac{(1 - a + b)^2}{2 - a + b}.$$

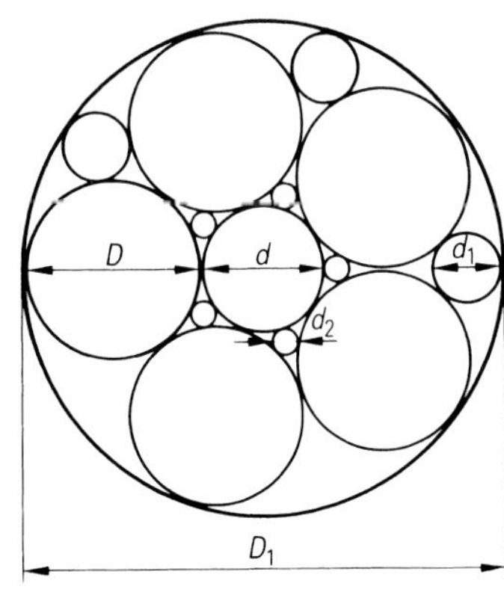

Figure 9.5
Schematic of a cable core

In Table 9.1 the values of these diameters for $c = 1$ to 12 stranding elements with $n = 2$ are listed.

The *slotted core cable* is a special design of the layer cable with one layer (Figure 9.6). In it the optical fibers lie in preshaped helical grooves or slots in the surface of the central member instead of being inside buffer tubes and stranded in layers.

Depending on the size and shape of these grooves in the central member, one or more fibers either individual or as a fiber ribbon or ribbon stack can move freely in them (Chapter 8.5). As with loose buffers, these grooves are usually filled with a compound. In cases where an unfilled cable design is required, water penetration can be prevented by use of water blocking tape or swelling yarn.

In order to further increase the fiber count of the cable, individual slotted core cable elements can be unit stranded inside a common outer sheath. In this design, cables with a fiber count of more than 100 optical fibers have the advantage of high packing density and simplified jointing technology due to the ordered position of the optical fibers.

The *maxitube cable* is another special example of optical cable design; it has no stranding elements stranded around a central member. Instead the optical fibers themselves are stranded inside a larger multifiber loose buffer, with this maxitube serving as the central member of the cable. Such a design is particularly simple and compact. It finds application, for example, in optical ground wires (OPGW) (Chapter 9.5.4).

There are two methods of stranding:
helical stranding and
reverse lay (SZ) stranding.

Table 9.1 Values of the various diameters for calculation of the cable core

Number of elements in layer	Total outer diameter D_1	Core diameter d	Outside interstice diameter d_1	Inside interstice diameter d_2
2	2.000 D	0.000 D	0.667 D	0.000 D
3	2.155 D	0.155 D	0.483 D	0.063 D
4	2.414 D	0.414 D	0.414 D	0.108 D
5	2.701 D	0.701 D	0.378 D	0.136 D
6	3.000 D	1.000 D	0.354 D	0.155 D
7	3.305 D	1.305 D	0.339 D	0.168 D
8	3.613 D	1.613 D	0.327 D	0.178 D
9	3.924 D	1.924 D	0.318 D	0.186 D
10	4.236 D	2.236 D	0.311 D	0.192 D
11	4.549 D	2.549 D	0.305 D	0.197 D
12	4.864 D	2.864 D	0.300 D	0.202 D

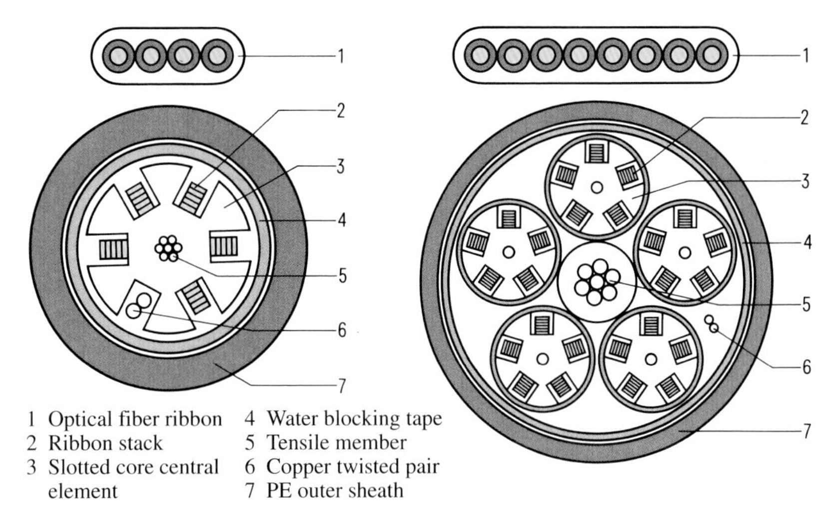

Figure 9.6 Slotted core cable with 100 and 1000 optical fibers

In *helical stranding* the stranding elements are stranded in one direction with a constant angle to the longitudinal axis of the cable.

In *reverse lay (SZ) stranding* the direction of stranding reverses after a predetermined number of revolutions, so that the stranding elements first describe an S along the cable axis and then, after reversing, a Z. At the reverse point they lie parallel to the axis of the cable (Figure 9.7).

Due to the stiffness of the stranding elements, a binder must be wound around the stranded elements in reverse lay stranding in order to retain them in the proper stranding position.

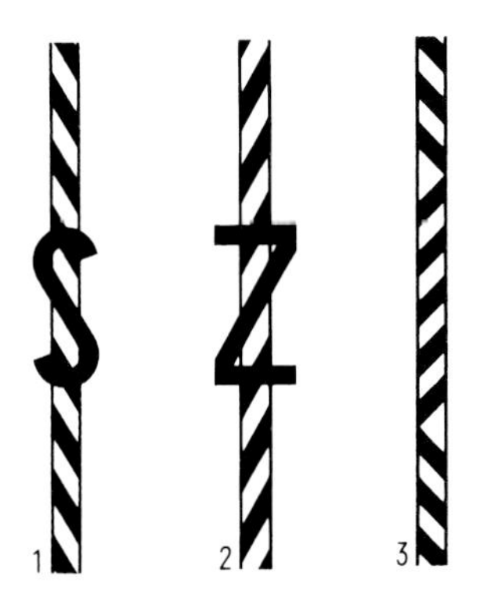

1 S direction
2 Z direction
3 SZ stranding

Figure 9.7
Reverse lay stranding

In helical stranding the stranding elements form a screw-type line comparable to a spiral staircase. Its length of pitch after a full turn of 360° is called the *lay length S*. The angle between the stranding elements and the cable cross-section is called the slope of stranding or *stranding angle* α. The distance between the axis of the cable and the middle of the stranding element is called the *stranding radius R*. Therefore the length L of a stranding element and the stranding angle α (Figure 9.8) can be calculated as

$$L = S\sqrt{1 + \left(\frac{2\pi R}{S}\right)^2};$$

$$\alpha = \arctan\frac{S}{2\pi R}$$

R	stranding radius in mm
L	length of stranding element in mm
S	lay length in mm
α	stranding angle in degrees
$2\pi R$	circumference of the stranding circle.

Because of stranding, the stranding elements must be longer than would be necessary if they were parallel to the longitudinal axis. The excess length due to stranding is given in per cent:

$$Z = \frac{L-S}{S}\cdot 100\% = \left\{\sqrt{1 + \left(\frac{2\pi R}{S}\right)^2} - 1\right\}\cdot 100\% =$$

$$= \left(\frac{1}{\sin\alpha} - 1\right)\cdot 100\%.$$

Z excess length due to stranding in %.

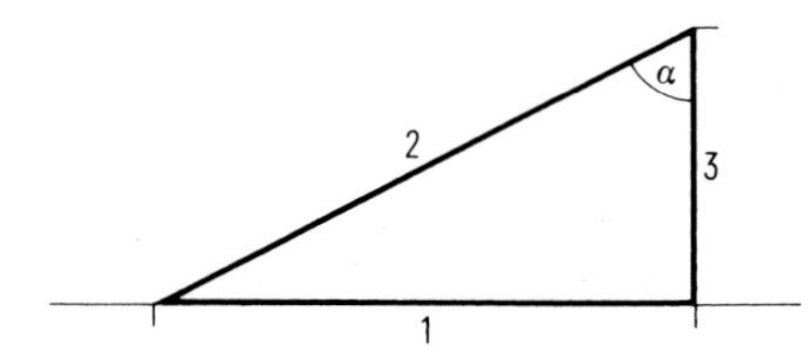

α Stranding angle α
1 Lay length S
2 Length L of stranding element
3 Circumference of the stranding circle $2\pi R$

Figure 9.8
Interdependence of lay length, stranding angle and the length of the stranding elements

The screw line or helix is a three-dimensional curve; its *radius of curvature* or *bending radius* can be calculated as

$$\varrho = R \cdot \left\{ 1 + \left(\frac{S}{2\pi R} \right)^2 \right\}$$

ϱ bending radius in mm.

For the strength and attenuation performance of an optical fiber it is important that it should not be bent too sharply. A measure of it is the minimum permissible radius ϱ. A typical value for standard fibers is $\varrho = 65$ mm. If the bending radius ϱ and the stranding radius R are both given, then the lay length is

$$S = 2\pi R \cdot \sqrt{\frac{\varrho}{R} - 1}.$$

Example

In an optical cable with helical stranding, a lay length $S = 102$ mm and a stranding radius $R = 4.3$ mm, the excess length due to stranding is calculated as

$$Z = \left\{ \sqrt{1 + \left(\frac{2\pi \cdot 4.3}{102} \right)^2} - 1 \right\} \cdot 100\% \approx 3.4\ \%.$$

Therefore for each 100 m of cable length the stranding elements are 3.4 m longer.

The stranding angle α is

$$\alpha = \arctan \frac{S}{2 \cdot \pi \cdot R} = \arctan \frac{102}{2 \cdot \pi \cdot 4.3} \approx 75.2°.$$

The corresponding bending radius ϱ is

$$\varrho = 4{,}3 \cdot \left\{ 1 + \left(\frac{102}{2 \cdot \pi \cdot 4.3} \right)^2 \right\} \text{mm} \approx 65.6 \text{ mm}.$$

In reverse lay stranding the bending radius along the cable core changes; it reaches a maximum at the reverse points and a minimum halfway between them.

In Figure 9.9 the bending radius ϱ as a function of the lay length S with a fixed stranding radius is shown for the two extreme cases of the reverse lay stranding and, for comparison, of the helical stranding.

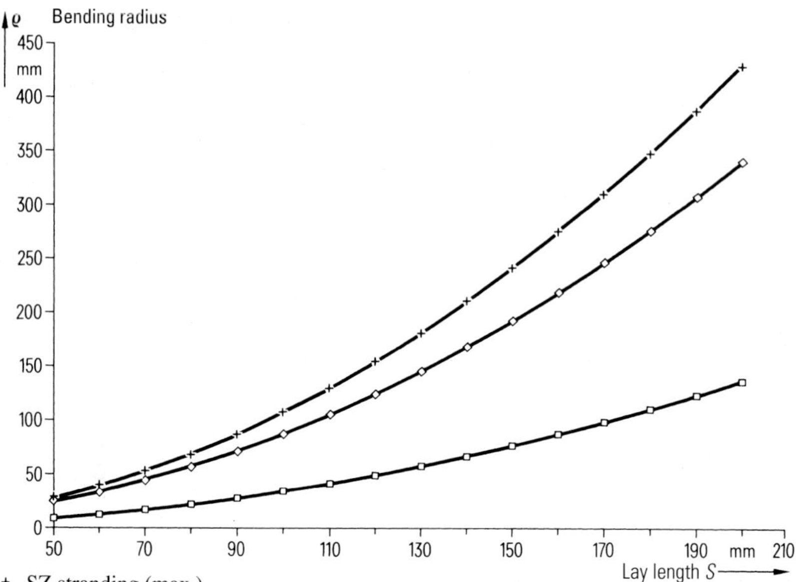

+ SZ stranding (max.)
◇ Helical stranding
□ SZ stranding (min.)

Figure 9.9 Lay length S as a function of the bending radius ϱ

In practical applications it is sufficient to use the equations for helical stranding for the reverse lay stranding.

In the following it is shown how the superposition of two bendings can be calculated. Such a case occurs, for example, in a unit cable, in which the stranding elements of the units have a bending radius ϱ_1 and the cable core is stranded again with a bending radius ϱ_2. The resulting total bending radius ϱ for the optical fibers is then

$$\frac{1}{\varrho} \leqq \frac{1}{\varrho_1} + \frac{1}{\varrho_2}.$$

9.1.2 Elongation and Contraction

In addition to the bending, the elongation and contraction of the optical fiber in the buffer tubes must be limited, so that for a specified tensile load and within a given temperature range for the optical cable no impermissible changes occur in the transmission parameters nor dangers to the optical fiber. It is known that in single-fiber and multifiber loose buffers the fibers can move freely inside their buffer tubes. In the neutral state (without stress) they are located in the center of the buffer tube. Their *clearance* ΔR is determined by the inside diameter d_i of the buffer tube and the outside diameter d_f of the optical fiber (Figure 9.10). In the multifiber loose buffer tube d_f must be taken to be the diameter of an imaginary circle which encircles the fibers as closely as possible.

The relative change in the length of an optical cable $\Delta L/L$, i.e. the permissible cable elongation ε_K or contraction ε_{TK} (contraction due to temperature), in an optical layer cable with a stranding radius R and lay length S is

$$\varepsilon = -1 + \sqrt{1 + \frac{4 \cdot \pi^2 \cdot R^2}{S^2} \cdot \left(\frac{2 \cdot \Delta R}{R} \pm \frac{\Delta R^2}{R^2} \right)},$$

where the plus sign in the expression in parentheses stands for *cable contraction* ε_{TK} and the minus sign for the *cable elongation* ε_K. This equation could lead to the conclusion that shortening the lay length would cause a major increase in the permissible elongation or contraction of the cable. However, the permissible bending radius must be taken into consideration, which is different for multimode and single-mode fibers (Chapter 9.1.1).

Example

In a typical multifiber loose buffer (inside diameter $d_i = 1.8$ mm) there are ten optical fibers (diameter $d_f = 4 \cdot 0.25$ mm $= 1.0$ mm) sharing a total clearance $\Delta R = (1.8 \text{ mm} - 1.0 \text{ mm})/2 = 0.4$ mm. With a lay length $S = 102$ mm and a stranding radius $R = 4.3$ mm the maximum permissible cable elongation is

$$\varepsilon_K = -1 + \sqrt{1 + \frac{4 \cdot \pi^2 \cdot 4.3^2}{102^2} \cdot \left(\frac{2 \cdot 0.4}{4.3} - \frac{0.4^2}{4.3^2} \right)} \approx 0.0062 = 6.2‰.$$

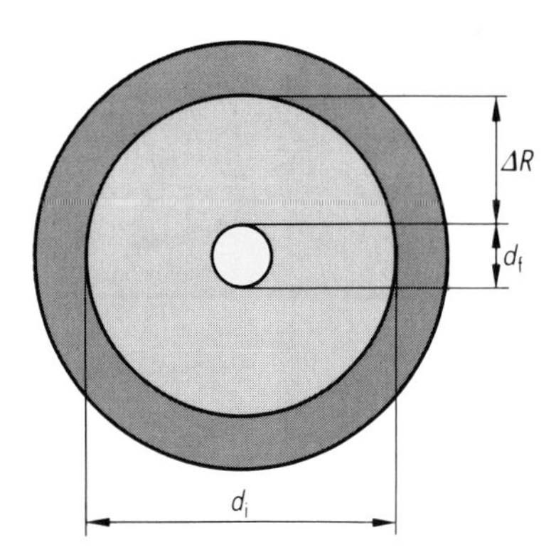

Figure 9.10 Buffer tube

To calculate the maximum permissible tensile force F_{max}, it is necessary to know the cross-sectional areas A_i of the materials in the cable and their Young's moduli E_i (Table 9.2). The sum of all the products $E_i A_i$ mulitplied by the maximum permissible cable elongation ε_K then gives the maximum permissible tensile force for the cable at which the optical fibers are not stressed:

$$F_{max} = \sum E_i \cdot A_i \cdot \varepsilon_K.$$

In practical applications it is sufficient to take into consideration only the materials in the central member and the strength member, which are stranded around buffered fibers and, together with the central member, assure the required tensile strength. This distributes the tensile forces evenly to the center and periphery of the cable core. The other materials used in the cable have a Young's modulus which is smaller by a power of ten, so that their cross-sectional area has little influence on F_{max}.

Table 9.2
Young's modulus, density and linear thermal expansion coefficient (typical values at room temperature)

	Young's modulus (N/mm²)	Density γ (g/cm³)	Linear thermal expansion coefficient α (1/K)
Fused silica glass	72 500	2.20	$5.5 \cdot 10^{-7}$
Thermoplastic polyester (PBTB = polybuthylen-therephtalat)	1600	1.31	$1.5 \cdot 10^{-4}$
Polyamide (PA)	1700	1.06	$7.8 \cdot 10^{-5}$
Polycarbonate	2300	1.20	$6.5 \cdot 10^{-5}$
Aramide yarn	100 000	1.45	$-2 \cdot 10^{-6}$
Spring steel for central member	200 000	7.85	$1.3 \cdot 10^{-5}$
LDPE (Low density PE)	200 to 300	0.910 to 0.925	
MDPE (medium density PE)	400 to 700	0.926 to 0.940	1 to $2.5 \cdot 10^{-4}$ (LDPE, MDPE, HDPE)
HDPE (high density PE)	1000	0.941 to 0.965	
Soft-PVC	60	1.31	$1.5 \cdot 10^{-4}$

A GRP (glass-reinforced plastic) element is used as the material for the central member in a fully dielectric optical cable;[1] it is composed of extremely strong glass filaments, bonded in an age-resistant and temperature-stable resin. In a non dielectric cable a steel element may also be used. If the diameter of an available GRP element or steel element is not sufficient for the cable core, then, for example, a firmly adhering PE protective overcoat can be applied to it to increase the diameter. Aramid yarns and/or glass elements are used as strength members. Such yarns provide a wrapping with high tensile strength which is both flexible and light.

In addition to the tensile force, which is generally only a factor during installation, an optical cable must also be able to withstand the effects of temperature changes over a specified range during its service life. In particular the shrinkage forces must be compensated for; these occur at low temperatures due to the contraction of the plastic materials in the buffer tubes and in the cable sheath.

For several materials, Table 9.2 lists the linear thermal expansion coefficients which describe the relative change in length ε_{TK} per temperature change ΔT:

$$\varepsilon_{TK} = \frac{\Delta L}{L} = \alpha \cdot \Delta T.$$

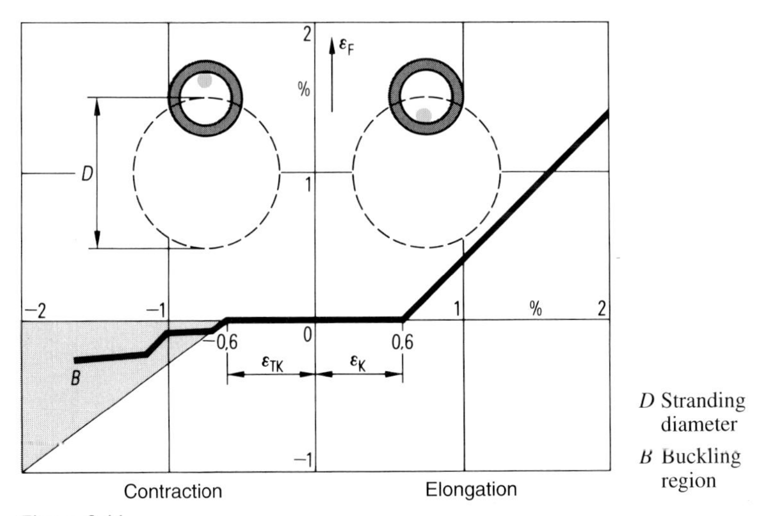

Figure 9.11
Optical fiber elongation ε_F, cable elongation ε_K and cable contraction ε_{TK} for stranded single-fiber loose buffers

[1]) cable without metallic elements

The relationship between elongation and contraction of a cable is shown once more in Figure 9.11 relative to the clearance of an optical fiber in a loose buffer. On the left the contraction is depicted, where the optical fiber touches the top of the inside of the buffer tube at approximately $\varepsilon_{TK} = 0.6$ per cent and, given greater contraction, i.e. a lower temperature, the fiber itself would be buckled ($\varepsilon_F < 0$). The right-hand side of the figure shows the reverse situation for elongation due to tensile forces. At about 0.6 per cent the optical fiber touches the lower inside of the buffer tube and is elongated itself only after a further increase in the tensile force.

9.1.3 Cable Core Filling

In order to assure longitudinal tightness of an optical cable when water enters it, the empty interstices in the cable core are filled with a compound at very high pressure (about 15 bar). This compound must have a composition such that it is not detrimental to the properties of the other cable elements. The compound has a negligible swelling effect on the PE sheath and a relatively low thermal expansion coefficient. A barrier layer of petroleum-resistant and relaxing[1]) thermoplastic adhesive (melt glue) extruded around the cable core serves on the one hand as an additional barrier for the filling compound and on the other hand as a seamless connection between the tension-resistant wrapping and the cable sheath, without detrimental effects to the flexibility of the optical cable.

If longitudinal water tightness is not required, as is usually the case for indoor cables, there is no need to fill the core. In order to protect the stranding elements in the cable core during the next production steps and in order to keep the strength members and sheath material out of the interstices, they are wrapped with one or several thin layers of plastic foil. As an identification of the source of the respective cable, i.e. to give the name of the manufacturer of an optical cable, a manufacturer's identification thread is applied during sheathing; this is usually parallel to the axis of the cable, directly over the stranding elements (e.g. in the order of green-white-red-white for Siemens and red-red-green-black for Corning Cable Systems).

At the request of the customer, it is possible to add a measuring tape. This is a paper strip 6 mm wide which has a continuous meter marking printed on it. In addition to the length markings printed on the cable sheath, this measuring tape provides a further means of identifying the length. The accuracy of this measuring tape as a measure for the length is about ± 0.3 per cent.

[1]) Relaxation: reduction of the elasticity.

9.2 Cable Sheath

The cable sheath should protect the optical cable core from

- ▷ mechanical,
- ▷ thermal and
- ▷ chemical effects,
- ▷ and additionally against moisture from the outside.

As with conventional cables, there are a number of sheath types from which to choose, depending on the conditions at hand. The popular and proven polyethylene (PE) (with or without a barrier layer, e.g. of aluminum) is used most frequently. In addition to polyvinyl chloride (PVC), used primarily for indoor cables, other sheath materials are available for special applications, e.g. perfluorethylene propylene (FEP), perfluoralkoxy copolymer (PFA) and ethylene vinylacetate (EVA).

If filled cables are required to have a sheath construction free of metals, then a barrier layer of plastic material such as polyamide (PA)-thermoplastic adhesive (melt glue) is added between the cable sheath and the yarns or the filling compound. It prevents the filling compound from migrating out of the cable core into the sheath.

This layer is applied in the same process step prior to extrusion of the sheath. Due to the thermoplastic adhesive (melt glue), a very tight connection of the cable sheath and the core with the aramid yarns is achieved. This proves very advantageous when the optical cables are pulled through a duct.

Theoretically, metal sheaths, e.g. lead or corrugated steel sheaths, can also be used over optical cable cores. However, it is necessary to take into consideration the fact that the tensile forces, which might occur when a cable is hauled through a duct, can cause permanent elongation of the cable and in the long run can affect the transmission properties of the optical fiber.

PE Sheath

Polyethylene (PE), a thermoplastic with the code 2Y (Appendix 15.2) and a polymerization product of ethylene, has the advantage of a low dielectric constant and a low loss factor. Due to its nonpolar behavior, it has largely constant dielectric properties over a wide range of temperatures; therefore it is used widely in electric cables. It meets all normal mechanical and chemical requirements and is particularly useful for outdoor cables.

There are kinds of PE for a wide variety of special applications of optical communications cables. In particular, due to their low density, high viscosity, break elongation and easy workability, they are used for sheathing optical cables cores. The main differentiations are low density PE (LDPE), medium density PE (MDPE) and high density PE (HDPE), Young's modulus, density and the linear thermal expansion coefficient can be found in Table 9.2

For heavy stresses on the cable sheath there are special PE materials available. They have a largely linear molecular structure and are produced with the help of catalysts in a new low pressure technique. These are linear low density PE (LLDPE) and linear medium density PE (LMDPE).

Sometimes the composite layer sheath or laminated aluminum PE (LAP) sheath known from conventional cable technology, is used. An aluminum band 0.2 mm in thickness is coated on both sides with an adhesive which adheres to plastic materials (copolymerisate). During production the outer layer is glued to the PE sheath due to the heat of the extrusion process. The inner layer serves as an adhesive in the overlap zone. The nominal thickness of the PE sheath (black) and the thickness of the PE in the composite layer sheath is usually 2.0 mm.

The PE sheath, which must lie tightly over the cable core, is applied in an extrusion process.

PVC Sheath

Polyvinyl chloride (PVC) sheaths of the type code Y (Appendix 15.2) are used mainly for indoor cables. The sheath usually has a gray color.

PVC is suitable for outdoor cables used in such places where aggressive deposits must be expected in the soil. For this purpose the PVC sheath mixture can be selected for oil resistance.

A PVC sheath is also advisable when flame resistance is required. For this case, however, it is important to note PVC is being replaced increasingly by halogen-free flame-resistant polymers (type code H).

Fluorine Plastic Sheath

The fluorine plastic material perfluorethylene propylene (FEP) with the type code 6Y (Appendix 15.2) is a viscoelastic material, as is PFA, and exhibits particularly good electrical, mechanical, thermal and chemical properties. Both are used for sheathing optical cables, especially when temperatures higher than 100 °C can occur. In those cases, care must be taken that the other cable elements including the optical fiber itself can withstand the anticipated temperatures. FEP and PFA sheaths have a smooth surface and are resistant to weathering.

Halogen-Free Sheath

Sheaths made of materials containing ethylene vinylacetate (EVA) with the type code H (Appendix 15.2) have a filling with up to 50 per cent by weight of aluminum trioxydhydrate. They are the preferred choice when both flame resistance and freedom from halogens are required. In these cases PE, PVC and even plastics with fluorine are not suitable.

Under the influence of flames at temperatures higher than 250 °C water separates out of the aluminum trioxydhydrate. By means of this separation and evaporation of water the flames lose energy. The temperature falls below that needed to maintain the fire and the concentration of inflammable gases and oxygen is reduced by the addition of steam, thus extinguishing the fire. The remaining material is incombustible aluminum oxide. Cables with EVA sheaths have mechanical properties almost identical to those of PE cables. The sheath is gray colored. As a more specific identification of the EVA sheath, FRNC (flame-retardant noncorrosive) is added to the end of the code for the cables (Chapter 9.6.4).

9.3 Protective Cover

PE and PVC and for special applications polyamide (PA) protective covers are used for outdoor and special cables; they protect the armoring over the sheath from corrosion and external damage, e.g. during ploughing in or during hauling in ducts.

Protective covers are usually produced from PE by the tube extrusion method at a temperature of about 200 °C and applied over bituminous (asphalt) corrosion-protecting materials.

When oil resistance is required or colored cable sheaths, protective PVC covers of the appropriate thickness are extruded over the PE sheaths. In all other applications black PE protective covers are used.

9.4 Armoring

Optical cables are often employed without armoring as buried and duct cables. The yarn strength members which are firmly connected to the cable sheath by the thermoplastic adhesive (melt glue), together with the antibuckling elements of the cable, are fully sufficient for the typical tensile forces.

For special applications, e.g. submarine cables or mine cables, for cables with rodent protection, for self-supporting aerial cables or cases in which extraordinary values for tensile stress and pressure must be compensated, an additional armoring is used to protect the optical cable core and the cable sheath.

Armoring elements must be found for these cables which do not greatly increase the weight of the cable or cause a major reduction in the flexibility and in addition exhibit a relatively low elongation; i.e. a good ratio between Young's modulus and the weight must be achieved.

Aramid yarns and steel in a wide variety of forms have become acceptable for optical cable design, whereby aramid yarns show a much higher ratio of Young's modulus and the weight than steel.

Aramid yarns are employed widely as tensile members under the various sheath materials. For higher tensile forces, as, for example, in self-supporting aerial cables, rovings of yarns are used under the sheath.

For cables with rodent protection, a corrugated steel laminated sheath may be used. In this case, a corrugated, corrosion resistant steel tape is applied lengthwise directly to the cable core and bonded to the PE sheath. The process is similar to that for the aluminum laminated sheath. The slight grooves in the steel tape improve the flexibility of the cable.

As an alternative to the corrugated steel laminated sheath, steel tape wrapping can be used. It consists of two layers of galvanized steel tape with a thickness of 0.1 mm each, which are embedded in the corrosion protection layers. An outer PE sheath is applied for protection.

Note that steel tape wrappings may not be used to counteract tensile forces.

For optical ground wire (OPGW) cables various materials have been accepted for the armoring with round wires. Depending on the application, one or more layers of galvanized steel or Aldrey (aluminum alloy – E-AlMgSi) wires are used and augmented with Stalum (aluminum clad steel (ACS), alumoweld) wires.

For other extreme applications, such as submarine cables, heavily galvanized sturdy round steel wires or stainless steel wires in one or more layers can be applied. Heavily galvanized steel wires still need additional reliable corrosion protection.

9.5 Construction Types

Optical cable designs are chosen according to specific characteristics and grouped into the following construction types:

outdoor cables,
indoor cables,
special cables.

As an explanation several typical cable types are selected, their applications described and existing specifications indicated.

In all construction types it is important to assure that neither during cable production nor through miscellaneous permissible influence are the transmission properties of the optical fibers permanently changed. It is possible to select the suitable optical fibers, with the exception of the dimensions of the buffer tube, independently of the basic construction type of the cable. Important criteria for the selection are the data of the route and the knowledge of the technical data of the planned transmission systems. Table 9.3 shows the values for the attenuation coefficient, bandwidth, etc., which generally apply to single-mode fibers and multimode fibers at 850 nm or 1300 nm and 1310/1550 nm.

Optical cables with copper conductors are marked on the accompanying documents, labels or on the drums with the CE sign (Conformité Européenne) in accordance with the low voltage EU Directive 73/23/EEC.

9.5.1 Outdoor Cables

Outdoor cables, which usually have a PE sheath, are constructed and dimensioned so that they can meet all requirements that might occur in buried or duct cable plants.

Depending on the number of optical fibers needed, either single-fiber loose buffers with one fiber or multifiber loose buffers with 2 to 12 fibers or maxitubes are used. Due to design and economical considerations, in cables with up to 10 fibers single-fiber loose buffers can be used while in those with more than 10 fibers, multifiber loose buffers or maxitubes are recommended. It is common practice that single-fiber and multifiber loose buffers or maxitubes are filled with a compound.

The following examples illustrate the advantages of multifiber loose buffers, especially in cables with many fibers.

	Outside diameter for single-fiber loose buffer design (mm)	Outside diameter for multifiber loose buffer design (mm)
Cables with 16 fibers	15	14.4
Cables with 120 fibers	24.5	22

In most applications, cables with single-mode fibers are designed with multifiber loose buffers or maxitubes.

Table 9.3
Transmission and optical properties of single-mode and multimode optical fibers at room temperature

		Multimode fibers G 50/1255		Single-mode fibers E 9/125	
		Wavelength 850 nm	Wavelength 1300 nm	Wavelength 1310 nm	Wavelength 1550 nm
Maximum attenuation coefficient [1])	$\frac{dB}{km}$	3.0 3.5	0.8 1.0 1.5	0.45	0.3
Minimum bandwidth for 1 km[1])	MHz	200 500 800 1000	200 500 800 1000 1200 1500	Some 10 GHz	
Maximum dispersion in the rqange of 1285 to 1330 mm at 1550 nm	$\frac{ps}{nm \cdot km}$	– –	– –	3.5 5 –	– 25
Nominal value for numerical aperture Permissible tolerance		0.20 ± 0.02	0.20 ± 0.02	– –	
Nominal value for mode field diameter Permissible tolerance	µm µm	– –	– –	9 ± 1	

[1]) According to ITU-T there are no exact specifications for attenuation and bandwidth categories. Only ranges, for example from 2.5 to 3.5 dB are specified.

Single-fiber loose buffers and multifiber loose buffers are both stranded in layers around a central member.

Cables with tight buffered fibers are usually not installed as outdoor cables.

A PE sheath or a PE laminated sheath is applied over the nonmetallic strength members. Due to the filling compound in the single-fiber or multifiber loose buffers and in the cable core, these outdoor cables with laminated sheath are also transversely watertight. In both cases the sheath is black.

Tensile forces between 1000 and 3000 N, such as those that typically occur during installation of outdoor cables in ducts, cause no problems if the correct cable design is chosen and proper cable grips or cap seals for the cable are used (Chapter 10.2). For particularly critical applications, an additional armoring with a protective cover over it is applied over the sheath. In cables for the Deutsche Telekom AG this is usually not the case. Trunk cables of the Deutsche Telekom AG are usually hauled into plastic ducts. Only in cases in which rodent protection is required and ducts cannot be considered an armoring consisting of a corrugated steel laminated sheath or as an alternative two steel tapes 0.1 mm in thickness with appropriate inner and outer protective covers are applied.

Cables usually carry a manufacturer's identification thread which has been trademarked as a mark of origin. These cables can have markings printed continuously on their sheaths or outer protective cover. The temperature ranges for these cables, both with regard to transportation, storage and operation and also during installation, can be found below.

Optical cables can be treated in the same way as copper cables during installation or hauling into ducts or shaped bricks, providing the minimum bending radius is taken into consideration. They even offer added advantage that, due to their low weight, lengths of 2000 m or more can be hauled into ducts.

Temperature Range

Transportation and storage temperature	– 40 °C to 70 °C
Installation temperature	– 5 °C to 50 °C
Operating temperature	– 30 °C to 70 °C

Standard delivery lengths of optical outdoor cables are 2000 m.

Depending on diameter, weight and applications, lengths of up to 5000 m or more can be produced.

Table 9.4 shows nominal values for several selected optical outdoor cables in comparison with copper cables.

Table 9.4 Comparison of optical and copper cables

Number of fibers		Outside diameter (mm)		Weight		Standard delivery lengths (m)	
Fiber	Cu twisted pair 0.6 mm diam.	Fiber	Cu	Fiber	Cu	Fiber	Cu
2, 4, 6	6	11.5	12	100	140	2000	1000
10	10	12	13.5	115	190	2000	1000
20	20	14.5	16.5	160	315	2000	1000
40	40	14.5	21.5	160	555	2000	1000
60	60	16	25.5	195	800	2000	1000
100	100	20	31.5	295	1245	2000	1000

In summary it can be said that for all applications of copper cable technology already known, adequate optical cables are available. Their major advantages are as follows:

▷ Mechanical

Smaller cable diameter,
lighter weight,
longer delivery length,
advantageous reel dimensions and therefore relatively low reel weight.

▷ Interference

Due to the possibility of designing a fully dielectric cable with corresponding materials, there are no interference problems from lightning, high voltage lines or electrical trains and no dc current conduction or grounding problems.

▷ Transmission

Good attenuation values:

single-mode fibers
at 1300 nm less than 0.45 dB/km
at 1550 nm less than 0.20 dB/km

multimode fibers with graded index profile
at 850 nm less than 2.5 dB/km
at 1300 nm less than 0.5 dB/km

High bandwidth:

for multimode fibers up to 1.2 GHz for 1 km
for single-mode fibers up to several times 10 GHz for 1 km.

▷ Relatively high protection against wire tapping

Figure 9.12 shows the comparison of the attenuation coefficients for different coaxial and optical cables, which is one of the main advantages of optical cables.

Optical cables are as suitable for systems with a small number of channels (< 30 channels), for transmission of telephone and remote control signals, and for planned broadband system of the future, such as broadband ISDN (Chapter 11.2.1), for which one network is used for all types of transmission, such as

> television, video text, radio/stereo,
> data, telex, telefax, facsimile,
> ISDN services, telephone and
> video conferences and videophone.

Repeater spacings can be achieved that are much longer than those for balanced and coaxial copper cables. This is already true today of digital systems with bit rates of up 2.5 Gbit/s.

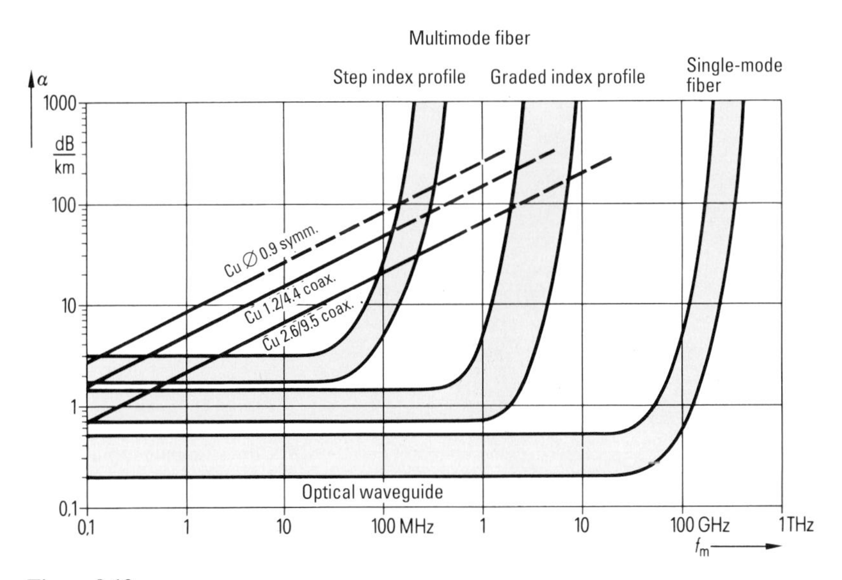

Figure 9.12
Attenuation coefficient as a function of the frequency of modulation for balanced and coaxial copper cables and for various optical cables

Outdoor Cables with Multifiber Loose Buffers for Single-Mode Fibers

In outdoor cables single-mode fibers are used predominantly. In constructing them, it is essential to take care that the optical fiber lies completely unstressed in the cable. Apart from that, all construction parameters which apply for multimode fibers also apply here.

Depending on the number of optical fibers, the following multifiber loose buffer is used:

2 to 12 fibers multifiber loose buffer 1.8 mm to 3.5 mm in diameter

Starting with the marking element, multifiber loose buffers are counted continuously. The copper pair or quad is not counted and its position does not specify any direction to be chosen for counting. Fillers are not counted.

In cables for the Deutsche Telekom AG, a pair or quad of copper wires with PE insulation is stranded instead of a multifiber loose buffer (Figure 9.13). Because multifiber loose buffer technology allows for flexibility in design, cables with high fiber count can easily be manufactured.

Cables with single-mode fibers are used in all cases where high bandwidth and low attenuation is required or long distances have to be covered with a minimum of regenerators.

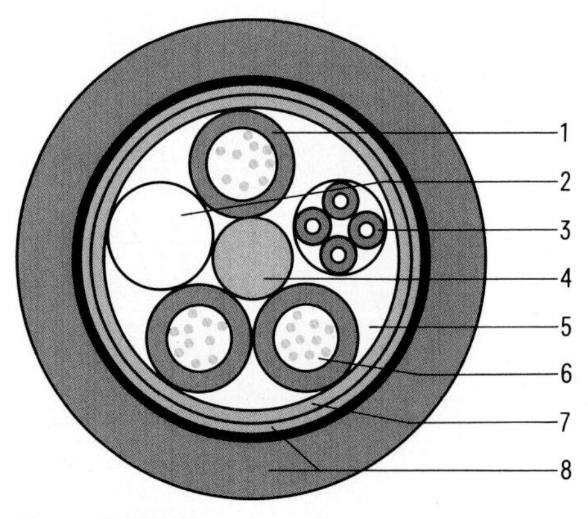

1 Multifiber loose buffer, counting element
2 Filler
3 Copper quad
4 Antibuckling member
5 Filling compound
6 Multifiber loose buffer
7 Core wrapping (optional)
8 Laminated sheath with strength members

Figure 9.13
Telecommunication cable of the Deutsche Telekom AG with a PE laminated sheath and 30 optical fibers in multifiber loose buffers and a copper quad

Code (Table 9.5)
A – DSF(L) (ZN)2Y 3 x 10E9 ... 9/125 ... F ... LG

Outdoor Cables with Multifiber Loose Buffers for Multimode Optical Fibers and Graded Index Profile

For outdoor cable design with graded index fibers multifiber loose buffers are normally used. This space-saving stranding element can hold up to 12 optical fibers and is filled with a compound. The color coding of the optical fibers can be found in Chapter 6.3, Table 6.3. The diameter of the multifiber loose buffers can be as large as 3.5 mm. At least five stranding elements are stranded in a layer around a central member. One of them can be planned as a spare multifiber loose buffer for 1 to 12 fibers that can also be replaced by a filler. Instead of a multifiber loose buffer a star quad of the same diameter with foam skin insulation may be stranded in for order wires.

The stranding elements can be coded as follows:

Spare multifiber loose buffer or filler substituted for it (marking element)	Red
Star quad (copper)	Yellow
Multifiber loose buffer	Green
Filler	Natural color

By stranding multifiber loose buffer very compact optical cables can be produced. There are standard cable designs with 20, 30, 40, 50, 60, 80, 100 or 120 optical fibers with 10 fibers in each multifiber loose buffer (Figure 9.14). For requirements of more than 120 optical fibers, there are also suitable cable designs available.

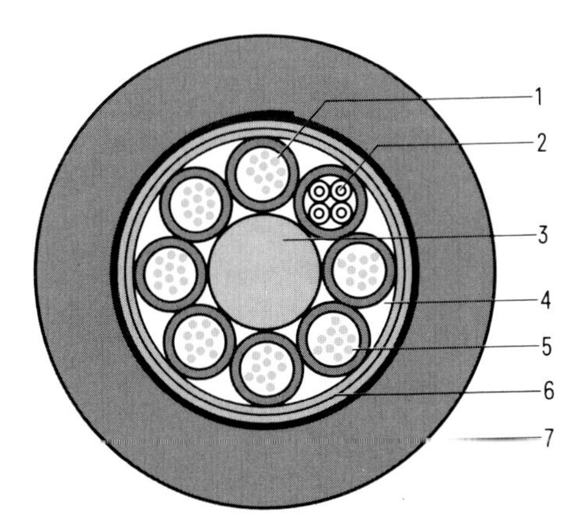

1 Spare multifiber loose buffer or filler
2 Copper quad
3 Antibuckling member
4 Filling compound
5 Multifiber loose buffer
6 Core wrapping
7 PE laminated sheath with strength members

Figure 9.14
Telecommunication cable with a PE laminated sheath and up to 60 optical fibers in multifiber loose buffers and one copper quad

Code

A – DSF(L) (ZN) 2Y 6 x 10G 50/125 ... F ... LG

Table 9.5
Explanation of codes for outdoor cables based on the German National Standard VDE 0888 Part 3

1 2 3 4 5 6 7 8 9[1])10 11 12 13[2]) 14

14: LG stranded in layers

13: Bandwidth in MHz for 1 km

12: Wavelength
B 850 nm
F 1310 nm
H 1550 nm

11: Attenuation coefficient (dB/km)

10: Cladding diameter (μm)

9: Core diameter (μm)

8: Type of fiber
E Single mode
G Graded index

7: Number of fibers or number of multifiber loose buffers × number of fibers per buffer tube

6:
B Armoring
BY Armoring with PVC protective cover
B2Y Armoring with PE protective cover

5:
2Y PE sheath
(L)2Y Laminated sheath
(ZN)2Y Laminated sheath with nonmetallic strength member
(ZN)(L)2Y Corrugated steel laminated sheath with nonmetallic strength member

4: F filling compound for filling the stranding interstices in the cable core

3: S Metal element in cable core

2:
H Single-fiber loose buffer, unfilled
W Single-fiber loose buffer, filled
B Multifiber loose buffer, unfilled
D Multifiber loose buffer, filled

1: Product designation
A-Outdoor cable

[1]) For single-mode fibers, the mode field diameter is given instead of the core diameter.
[2]) For single-mode fibers, the dispersion is given in ps/nm · km

All other structural elements, such as sheath, armoring, protective cover, etc., are used corresponding to the diameter in question.

In comparison with coaxial cables, optical cables weigh much less.

For example:

Coaxial cable type 32 c of the Deutsche Telekom AG weighs approximately 5 kg/m, as opposed to a comparable trunk cable with 60 optical fibers which weighs approximately 350 g/m.

9.5.2 Indoor Cables

Optical indoor cables are needed for a wide variety of applications inside buildings, because outdoor cables may only be used in a limited way inside buildings. Therefore they are terminated just inside the premises in an optical fiber distribution closure or in a cable termination rack. In order to assure that the splice joints show only low attenuation increases, the indoor cable should have similar optical transmission properties as the outdoor cable to which it corresponds. Indoor cables are also designed such that mechanical stresses or temperature variations will not negatively effect their properties.

In general three indoor cable types can be distinguished:

Multi-indoor cables (MIC)
Break-out cables (Fan-out)
Interconnect cables

All indoor cables are free of metal. For that reason they require no grounding and no lightning protection. The all-dielectric construction insures EMI (electromagnetic immunity). All indoor cable types consist of 900 μm TBII tight buffered fibers which are very easy to strip (Chapter 8.3). They are stranded differently depending on the type of cable. Fiber type and length markings can be printed on the outer sheath for easy identification.

In addition to the standard design for indoor cables, other constructions are possible to meet particular specifications (Table 9.6).

Multi-Indoor Cables (MIC)

Multi-indoor cables are rugged, high performance optical communication cables. They are designed for various indoor and outdoor requirements, including routing between buildings within ducts, inside buildings up riser shafts, in plenum spaces and for fiber-to-the-desk (Chapter 11.5).

Table 9.6
Explanation of the codes for indoor cables based on the German National Standard VDE 0888

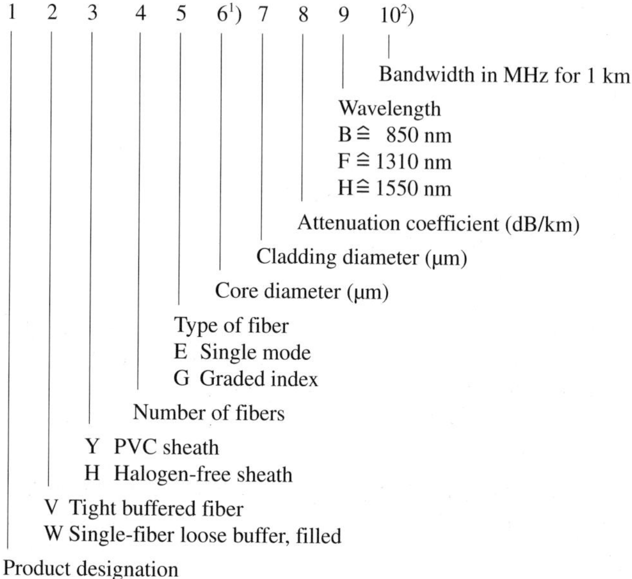

[1]) For single-mode fibers the mode field diameter is given instead of the core diameter.
[2]) For single-mode fibers the dispersion is given in ps/nm · km.

Available in a wide range of fiber counts, this cable family provides the necessary bandwidth capacity to transport all voice, data, video and imaging signals normally required in today's evolving office and factory environment. Also the cable's small diameter, light weight and flexibility allow for easy installation, maintenance and administration, particular in space-constrained areas.

Multi-indoor cables consist of stranded tight buffered fibers and nonmetallic strength members (e.g. aramid yarn). In cables with more than 6 optical fibers, the fibers are stranded around a nonmetallic central member to provide a more robust cable (Figure 9.15).

Cables with a fiber count of more than 24 are organized in subunits for configuration control, ease-of-installation and quick repair. In such a cable 6

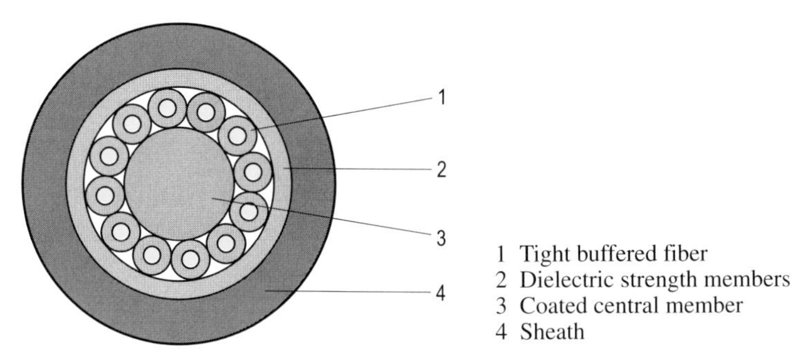

Figure 9.15 Multi-indoor cable with 12 optical fibers

tight buffered fibers each are stranded within a subunit with a nonmetallic strength member. The subunits themselves are stranded around a nonmetallic central member (Figure 9.16). For cables with a fiber count of more than 72 fibers, the subunits consist of 12 optical fibers. For multi-indoor cables a fiber count of 2 to 144 is standard.

The maximum long term tensile load for MIC is between 200 N for a fiber MIC and 4000 N for a 144 fiber MIC. For short term the permissible tensile load is four times higher.

The individual fibers are color-coded for easy identification.

Break-Out Cables (Fan-Out)

Break-out cables enable the individual routing or fanning of individual fibers for termination and maintenance. They come in three different constructions: 2.7 mm subunits for heavy duty applications, 2.4 mm subunits for standard duty applications and 2.0 mm subunits for light duty applications (Figure 9.17). These cables can be routed between buildings within ducts below the frost line, inside buildings up the riser shafts, under computer room floors and in fiber-to-the-desk applications.

However, due to the significant higher costs of break-out cables compared to MIC as well as the higher weight and size of these cables the main application field should be short and medium transmission distances. In most cases MIC cables are sufficient for the installation. Only in environments with very rough conditions or the real need of routing single fibers individually, break-out cables bring in the advantage of their construction.

The subunits can be numbered or color-coded for easy identification. The standard fiber count for break-out cables is between 2 and 24. The maximum

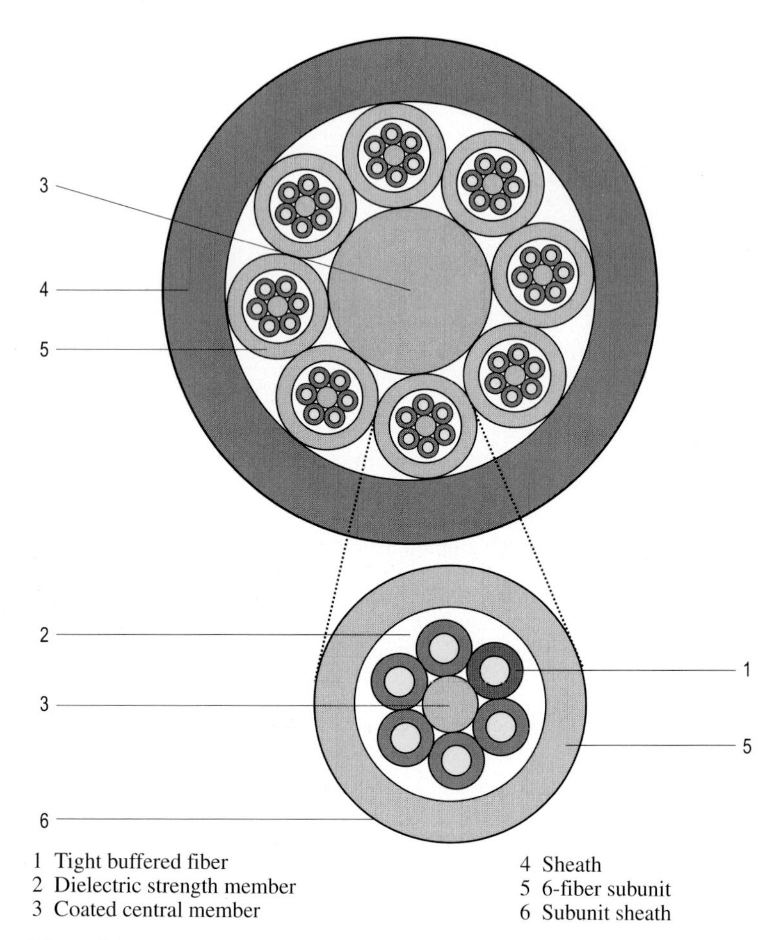

1 Tight buffered fiber
2 Dielectric strength member
3 Coated central member
4 Sheath
5 6-fiber subunit
6 Subunit sheath

Figure 9.16 Multi-indoor cable with 48 optical fibers in 6-fiber subunits

long term tensile load for break-out cables is between 300 N for a 2 fiber break-out and 1600 N for a 24 fiber break-out. For short term the permissable tensile load is three times higher.

Interconnect Cables

Cables for interconnecting equipment are designed for voice, data, video and imaging transmission in computer, process control, data entry and wired office systems. Available in one- and two-fiber designs, these cables are optimized for ease of connectorization and used as jumpers in intrabuilding distribution (Figures 9.18a, b, c).

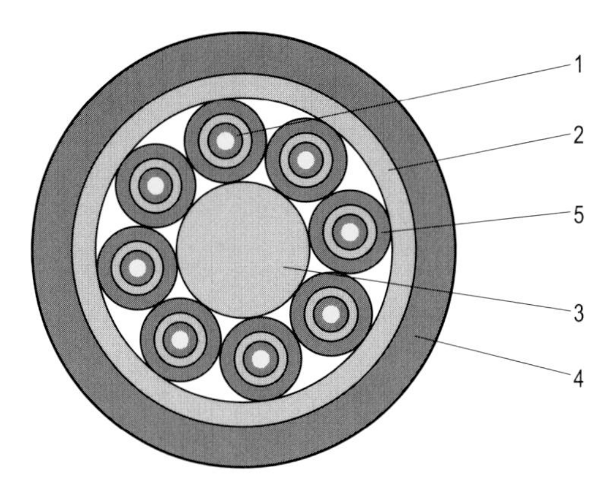

1 Tight buffered fiber
2 Dielectric strength member
3 Coated central member
4 Sheath
5 1-fiber subunit

Figure 9.17 Break-out cable with 8 optical fibers

Small diameter and bend radius provide easy installation in space-constrained areas. They allow easy direct connectorization or come preconnectorized from the factory as cable assemblies for use in work areas or as cross-connect patch cords.

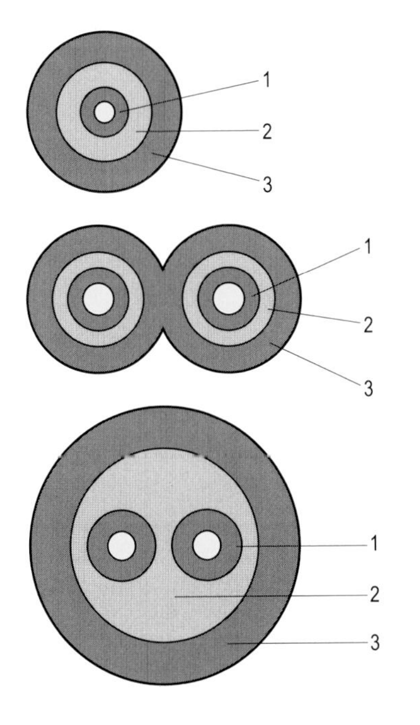

1 Tight buffered fiber
2 Dielectric strength member
3 Sheath

Figure 9.18 a
Interconnect cable with one optical fiber

Figure 9.18 b
Interconnect cable with two optical fibers, Zipcord

Figure 9.18 c
Interconnect cable with two optical fibers, DIB (**d**ual **i**ntra**b**uilding cable)

9.5.3 Indoor-Outdoor Cables

Indoor-outdoor cables are fully waterblocked for indoor and outdoor applications. These cable types eliminate the need for transitional splicing or direct termination when entering a building. These cables can be used for premises drop out to the first outdoor splice point or for high fiber count indoor trunking, especially in areas with limited conduit or vault space, as these cables can have fiber counts of up to 288 and higher (Figure 9.19). They must meet the specific fire codes in the country of application.

In order to reduce the fire risk, flame retardant materials are used for the sheath and tape. The fibers are placed in a standard loose buffer tube (e.g. 12 fibers each) with filling compound, while the buffer tubes themselves are stranded around a dielectric central member and wrapped with waterblocking tapes. Additional waterblocking threads are used in the unfilled cable core.

A major application of indoor-outdoor cables is the connection of a main distribution frame (MDF) in a building directly from the outside. No transitional splicing between outside cable plant and MDF is required.

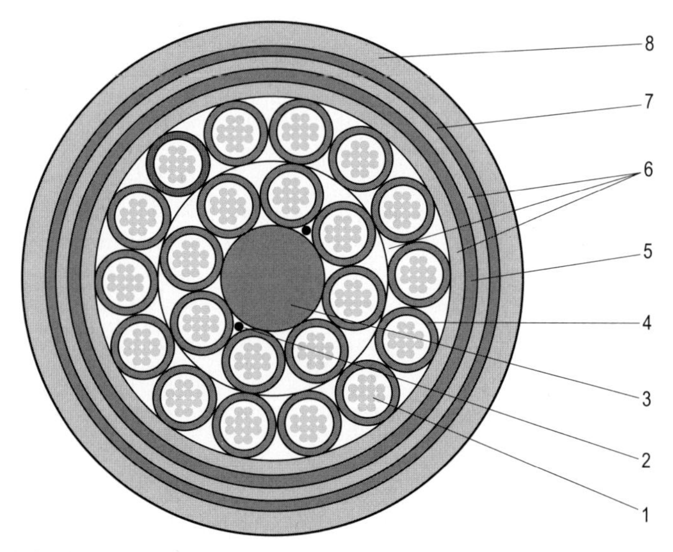

1 Optical fibers
2 Waterblocking threads
3 Dielectric central member
4 Loose buffer tube (12 fibers)
5 Flame retardant tape
6 Waterblocking tape
7 Dielectric strength member
8 UV-resistent, flame retardant outer sheath

Figure 9.19 Indoor-outdoor cable with 264 optical fibers

Application of Indoor and Outdoor Cables

The following example, shown in Figure 9.20, gives an overview of the application of different indoor and outdoor cable types: Two buildings (B1 and B2) on a campus environment need to be connected in the *Outdoor Area*.

In building B1 the main distribution frame (MDF) is located in the basement to handle the distribution to each floor. From the MDF all floors of the building will be connected. The vertical area in the building is called "Riser Area". On each floor an intermediate distribution frame (IDF) is located to handle the distribution on that floor. This horizontal area is called *Plenum Area*. Finally the area from the telecommunication outlet (TCO) at the wall to the workstation is called the *Patch Area*.

In building B2 the area to be wired is only the upper floor. For that reason the MDF is also the IDF and located in the same floor as the workstations. Here only the plenum area needs to be cabled.

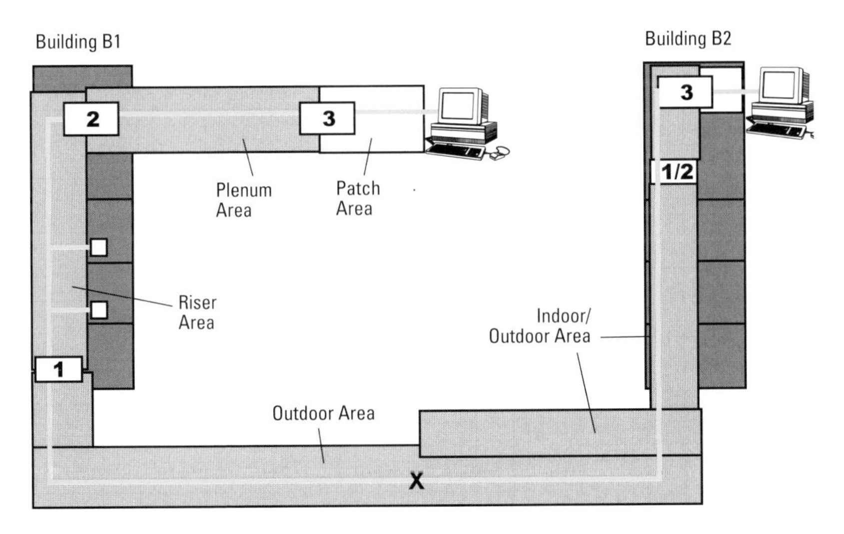

X Splice Point
1 Main distribution frame (MDF)
2 Intermediate distribution frame (IDF)
3 Telecommunication outlet (TCO)

Figure 9.20 Campus cabling

The workstations in the patch area will be connected by interconnect cables. In most cases two fiber zipcord cables are used. In case of IBM Escon interfaces at the workstations, DIB cables will be used.

The plenum area will be covered in both buildings with multi-indoor cables (MIC) with a fiber count of 4 to 8 fibers, depending on the number of telecommunication outlets (TCO) (two fibers per outlet are required). Each outlet can serve one workstation or printer. Therefore, in general (number of TCOs divided by 2 or 4) cables will be necessary per floor. Alternatively in very rugged environment, break-out cables with the same fiber count can be used. However, these cables are more expensive, heavier and larger in diameter. The cables will be terminated at the IDF.

In building B1 for the riser area a high fiber count MIC will be used. From the main-distribution frame a MIC will run to each floor. These cables will be terminated at the IDF. The number of fibers per cable should be at least twice the number of planned TCOs per floor.

At the basement in building B1 the MIC will be terminated at the MDF. From the MDF outdoor cables will run through the campus. At a splice point in the outside area – normally located in a manhole – an indoor-outdoor cable will be spliced to the outdoor cable. This indoor-outdoor cable will run the last few meters outside to building B2 and inside the building all the way up to the upper floor. Here the cable will bc tcrminated at the MDF/IDF. From there the TCOs are again connected by 4 to 8 fiber MIC or break-out cables. The number of fibers in the indoor-outdoor cable should be at least twice the number of planned TCOs on the floor.

With the use of the indoor-outdoor cable a transition point inside the building is omitted. Due to different fire codes, in many countries it is not allowed to use standard outdoor cables inside a building. Therefore, usually an outdoor cable must be terminated in the basement of a building and then an indoor cable must be used in the riser and plenum area.

The indoor-outdoor cable is allowed to be used inside as well. Such a cable can run from the outside all the way up to the MDF/IDF. There the cable will be terminated and a MIC or break-out cable will be used in the plenum area.

9.5.4 Aerial Cables

The worldwide demand for fiber optic aerial cables has increased, especially for use in high voltage power systems as well as railway cable networks. The insensitivity of fiber optics to electrical and magnetic interferences makes these cables an ideal transmission medium for both telecommunication, control and measurement signals.

The trend is towards suspending all types of cables from poles or pylons, weight permitting, whenever it is not declared a necessity to bury them beneath the ground. This does, however, involve risks. The cable must for example be installed so that the cable's tensile strength can withstand its own weight, the effects of wind and the weight of ice. Physical damage, such as that from falling trees and shotgun fire, must also be prevented by means of suitable design measures.

In most countries overhead cableways are already in place, e.g. extra-high voltage cables and railway cables. This existing infrastructure can be used by replacing ground wires by **op**tical **g**round **w**ires (OPGW) or by integrating **a**ll **d**ielectric **s**elf-**s**upporting cables (ADSS) in high voltage transmission lines.

Fiber Optic Cables in Overhead Ground Wires (OPGW)

Self-supporting fiber optic cables in overhead ground wires have been used as a substitute for traditional overhead ground wire cables by power supply companies in Germany since 1986. A prerequisite is that the OPGW does not differ significantly from traditional ground wire cables in either weight or diameter, so that the load on the suspension poles remains the same.

The optical ground wire cables come in two basic designs. Optical fibers are either embedded in a central tube or in stranded multifiber metal tubes. When bundled in a central tube, armoring consists of either a double (Figure 9.21a and b) or single (Figure 9.21c) layer of wires. When the optical fibers are in stranded multifiber metal tubes, these tubes replace one or more of the armoring wires in the inner layer (Figure 9.21d).

Typically the core of double armored cables consists of a plastic or metallic (high grade steel) multifiber tube. The inner layer of armoring is usually made up of galvanized steel wires or aluminum-coated steel wires (AW – alumoclad, alumoweld). Aldrey wires (AY) usually make up the outer layer of armoring. Aldrey, a highly conductive alloy of aluminum, magnesium and silicon, has twice the tensile strength of pure aluminum.

Depending on the number of fibers, the diameter of the plastic tube can vary from 3.5 mm to 8 mm while the metallic tubes can be up to 6 mm in diameter. Tests have been carried out on cables containing as many as 96 fibers in one tube.

In cables with stranded multifiber metal tubes the tubes are about the same diameter as the armoring wires (an outer diameter of about 2.3 mm to 3.6 mm). As many as 16 to 36 fibers can be inserted in each tube.

The basic design philosophy is to prevent any residual strain from the optical fibers. Therefore the wires' cross-sectional area, their quantity and the cross-

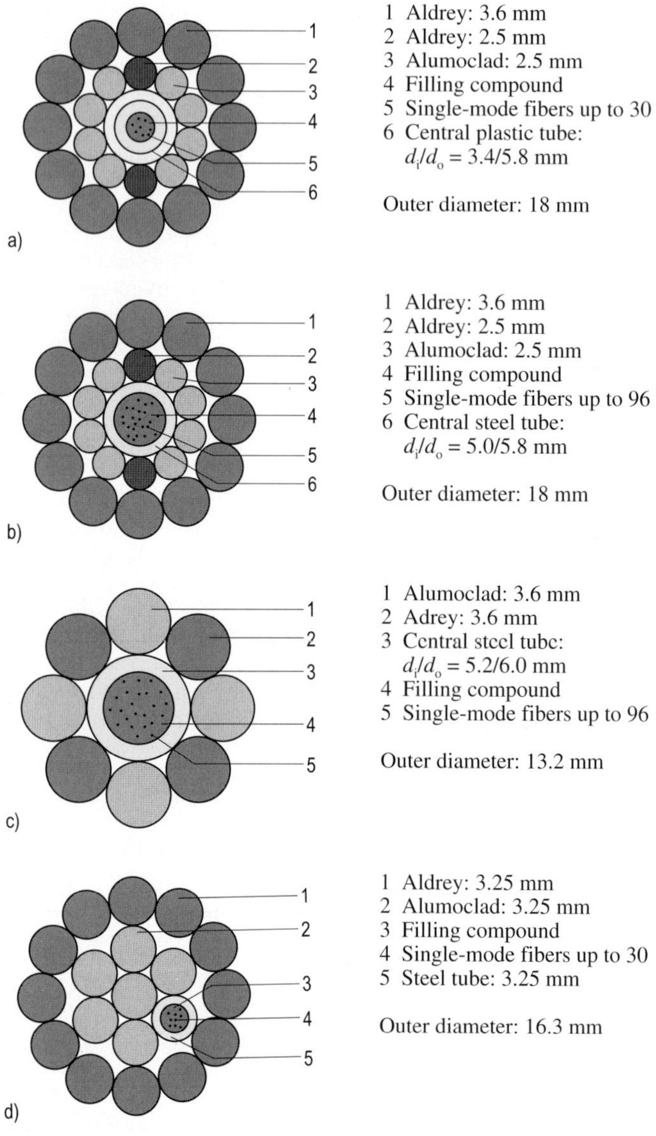

Figure 9.21
OPGW designs

a) Central plastic tube b) Central metallic tube
c) Central tube with one armoring layer d) Stranded stainless steel tube

sectional relationship at the two wire types are defined with respect to mechanical tensile strength and current-carrying characteristics. Arranged in a helical formation the fibers are not exposed to any stress even under severe wind and ice loads. A typical excess fiber length is 0.5% (Chapter 8.2.1).

In order to counteract the thermal loading under short-circuit conditions, all constructional elements must be suited to one another. Therefore the outer armoring layer should consist mainly of aldrey wires, which are superior conductors. In this way it is possible to keep the temperature of the inner armoring layer so low that the plastic compounds of the central tube (tube material, fillers, polypropylene buffer) are not damaged.

The specified nominal short-circuit current (1 s duration) is determined by the maximum allowable temperature increase (140 °C) of the outer armoring layer. At the upper limit aluminum and aldrey begin to loose their tensile strength. For example, for a cable with 112 mm^2 aldrey and 32 mm^2 aluminum-coated steel wires the temperature increase after a short-circuit with a short-circuit current of 11.4 kA for one second is 123 °C in the outer aldrey layer and only 73 °C inside the plastic buffer tube. Electrical loads as high as two times that of the specified value do not lead to any change in the optical transmission properties.

Long term performance studies have been carried out, which assess the vibration effects caused by wind in combination with a salt-fog. After a test period of approximately three months when the cables were finally examined, no breaks were found, and no changes in attenuation recorded. The steel tubes were not corroded. Only slight fretting marks were visible on the surface of the tubes. Plastic tubes were not affected by this test.

All Dielectric Self-supporting Cables (ADSS)

ADSS cables are mainly used to upgrade existing high voltage power lines with fiber optics without replacement of the groundwire. Their all dielectric design minimizes the danger during installation and prevents short-circuits in case of contact with the phase wires.

The cross-section of a typical ADSS cable is shown in Figure 9.22. The core of the cable is formed by a central dual-layer maxitube in which up to 48 fibers can be integrated. The fibers are inserted with a specific excess length. They are therefore not subjected to any stress if the cable is loaded with a specified tensile force. In order to prevent water penetration and migration the tube is jelly filled. Aramid yarns are wrapped around the core to provide the required tensile strength.

For installation on power lines operating at 110 kV and above, the outer jacket is made of special polymer, which is tracking-resistant and self-extin-

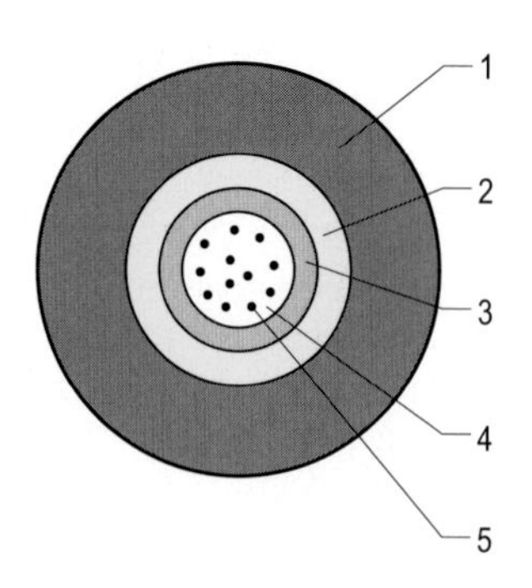

1 PE sheath
2 Aramid yarns, strength members
3 Buffer tube
4 Filling compound
5 Optical fibers

Figure 9.22
All dielectric self-supporting cable

guishing. The aramid yarns are treated with neutral ionogenic fluid to absorb the capacitive currents and eliminate any electrical stresses from the surface of the outer jacket. In the case of overhead power lines operating below 110 kV, the jacket material is polyethylene (PE).

The maximum span length mainly depends on the amount of strength members to be used (e.g. aramid yarns) and the required sag. With 15 mm^2 aramid yarns a typical span length is in the order of 300 m.

All Dielectric Lash Cables (AD-Lash)

Another class of fully dielectric cables are the AD-Lash cables. They are specially designed to be lashed on ground or phase wires (Figure 9.23). This cable consists of a dual layer, jelly filled maxitube with up to 48 fibers which is surrounded by a polyethylene sheath (Figure 9.24).

To provide a certain tensile strength, which is mainly required during the installation procedure, two thin aramid yarns are embedded in the outer PE sheath. A typical outer diameter of the complete cable is 5 to 6 mm. Therefore the weight per km is only about 30 kg. Such a cable is ideal for the use in transmission lines where the poles are not strong enough to withstand the additional tensile load caused e.g. by a self-supporting cable such as ADSS.

By wrapping an adhesive tape around both the cable and the ground or phase wire they are lashed together. This is done with a special machine which moves on the ground or phase wire. It consists of a rotating tape pay-off and the cable pay-off reel with approximately 2 to 4 km of AD-Lash cable.

Figure-8 Cables and Multipurpose Cables

Besides the above mentioned aerial cables for special application in high voltage power transmission lines aerial cables for general applications are available.

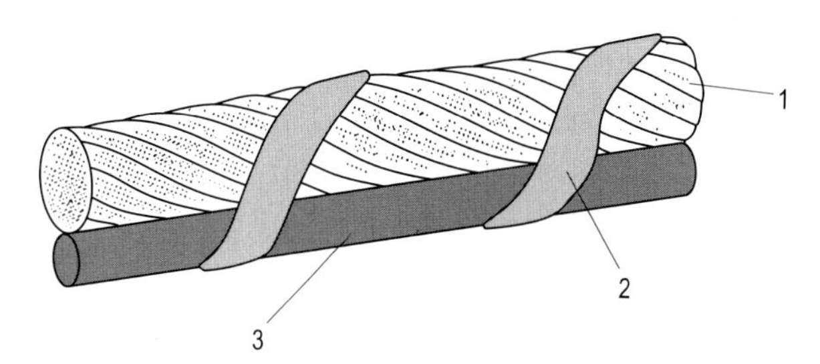

1 Ground wire
2 Adhesive tape
3 Fiber optic cable

Figure 9.23 Lashing of the AD-Lash cable to a ground wire

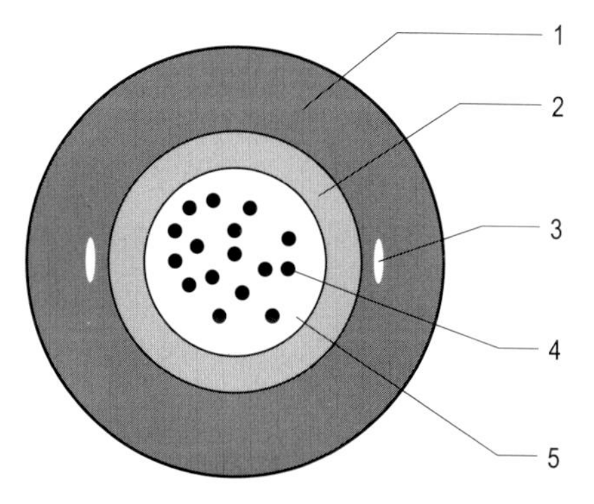

1 PE sheath
2 Buffer tube
3 Aramid yarns, strength members
4 optical fibers
5 Filling compound

Figure 9.24 AD-Lash cable

Figure 9.25 depicts a cross-sectional diagram of a so called figure-eight design. It consists of two portions: The upper half contains a strain-bearing element (messenger wire) consisting of aramid yarns or a steel rope. The lower half contains the cable's core made up of a central buffer tube or stranded tubes. Both halves are embedded in a polyethylene sheath. This design is mainly used for short pole distances in the order of 50 m.

A multipurpose cable (Figure 9.26) which also can be used as a self-supporting aerial cable consists of a maxitube (Chapter 9.1.1) as the essential component, which can accommodate up to 96 fibers. A polyethylene sheath, in which two oppositely-placed strength members are embedded, is extruded directly onto the maxitube. This cable is very light and is perfectly suited for weak poles up to a distance of approximately 50 m, depending on the acceptable amount of sag. If aramid yarns are used instead of steel wires for the strength members, this cable may be used as a fully dielectric aerial cable in regions highly prone to lightning strikes.

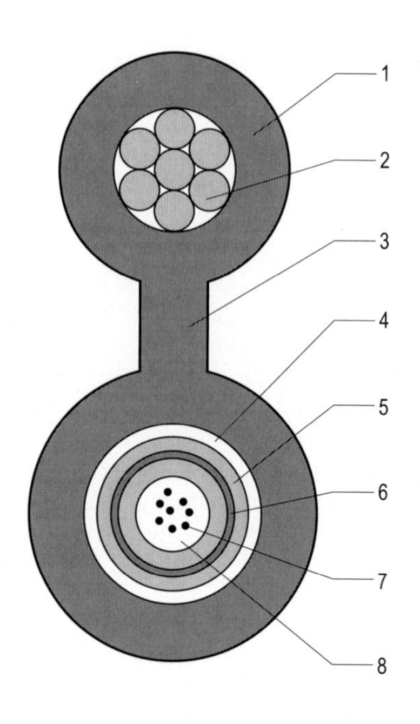

1 PE-sheath
2 Steel-rope
3 Spacer
4 Corrugated steel tape
5 Swellable tape
6 Buffer tube
7 Optical fibers
8 Filling compound

Figure 9.25
Figure-8 cable

9.5.5 Submarine Cables

To extend terrestrial type fiber optic transmission capabilities to repeaterless undersea applications the MINISUB unrepeatered submarine cable product line has been developed. MINISUB equipment consists of standard off-the-shelf terrestrial terminals and maintenance systems, optical boosters and

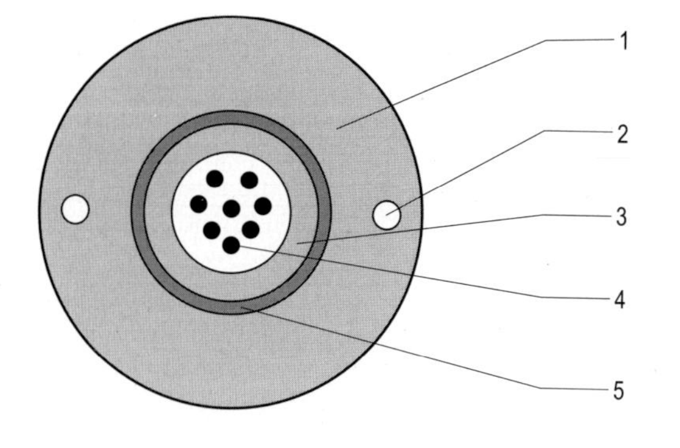

1 PE-sheath
2 Strength members
3 Buffer tube
4 Optical fibers
5 Yarn spinning

Figure 9.26
Multipurpose cable

preamplifier modules as well as low loss optical fibers and high performance, mechanically robust and low cost cables that are specially tailored for unrepeatered system applications.

Cable laying and recovery in sea depths down to 6000 m, total unrepeatered span length of 400 km with a transmission capacity of up to 24 fibers with 2.5 Gbit/s in SDH technology (Chapter 14.1) are key features.

One fundamental component of the MINISUB system is the submarine cable and its accessories. The MINISUB cable family is based on the central loose buffer tube design (Figure 9.27). To withstand the high hydrostatic pressure at deep water depths and to seal the optical fibers hermetically against molecular hydrogen, the buffer tube is made up of copper. Depending on the expected tensile load during laying, operation and recovery, the armoring consists of one or up to three layers of high tensile strength galvanized steel wires. The armoring layers are typically separated by polyethylene (PE) layers. The outer layer also consists of PE.

For protection of the cables from fishing activities as well as anchoring in coastal regions the cables are buried one meter deep in the sea bed with the help of a specially designed underwater plough. Another possibility for the improvement of the tensile behavior of the cables is the use of a so-called protector. The add-on protector consists of two steel ropes which can be manufactured independently from the MINISUB cable in the cable factory and later on attached to the cable on board of the laying ship.

9.6 Environmental Tests on Optical Cables

A particularly wide range of requirements are placed on optical cables with regard to transmission properties and environmental conditions. Whereas the transmission properties depend mainly on the characteristics of the optical fiber (the test and measurement methods for the most important transmission parameters are discussed in Chapter 5), the response of the cable to environmental influences is dependent on the cable design. There are a number of mechanical and thermal test methods that can be used to optimize the cable properties to suit the respective applications. With an exact evaluation of the results, it is possible to predict the performance of the optical cables under the expected operating conditions. The most important tests are for tensile strength and temperature.

The environmental test methods are specified in European standard EN 60794-1-2 (national German standard VDE 0888 Part 100) and internationally in IEC 60794-1-2.

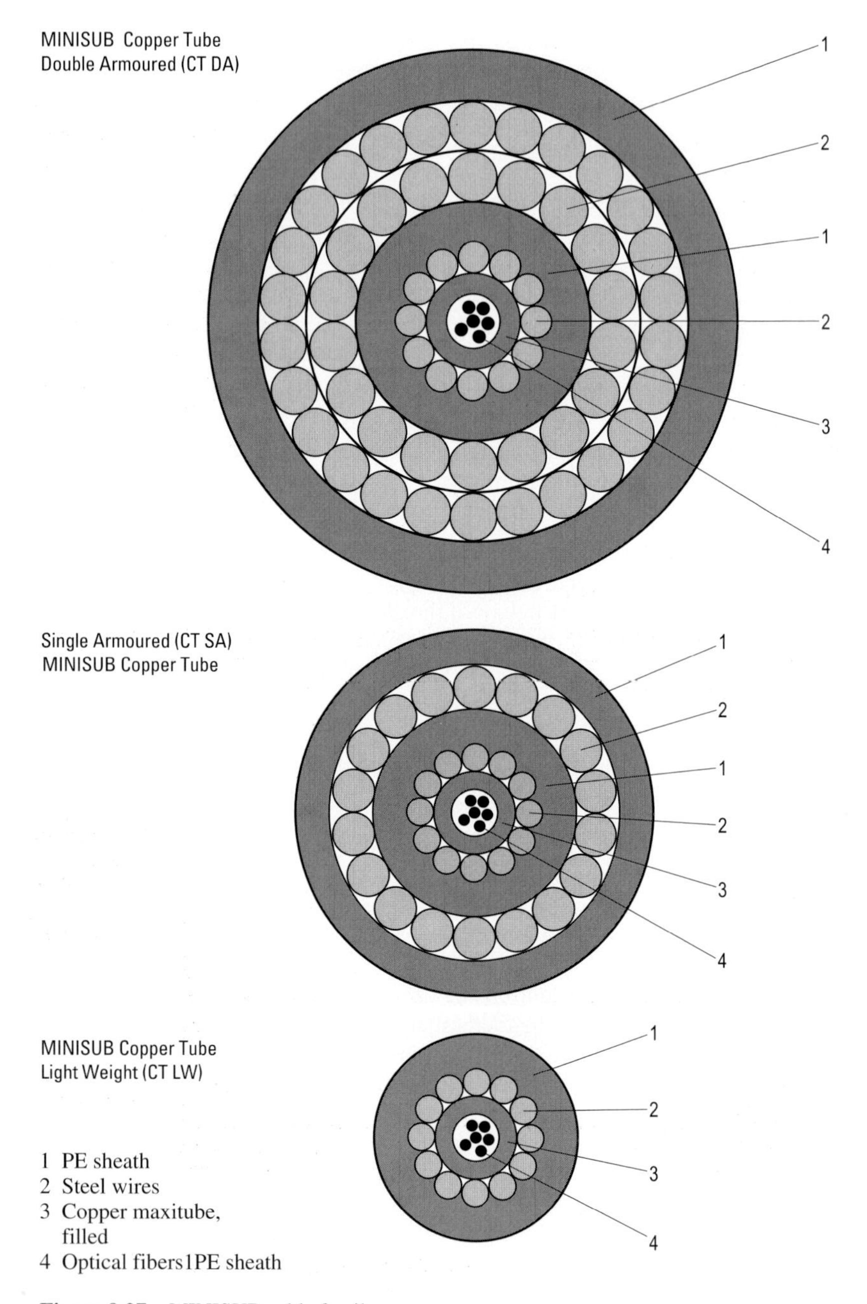

Figure 9.27 MINISUB cable family

9.6.1 Test of the Tensile Strength

Strong tensile stresses can occur during cable installation; they must not affect the functioning of the optical fiber. This is also true of installed cables in which residual tensile forces must not have any permanent effect on the optical fiber.

For this reason single-fiber and multifiber loose buffers are mainly used in optical cable design (Chapters 8.1 and 8.2). The fibers have freedom of movement in these loose buffers, which keeps them decoupled from any major stress outside. This free space is the operating window of the optical fiber (Figure 9.28); depending on the specification, it is designed and produced and then tested during the course of a type test on the finished cable. During type testing the range of elongation is determined in a tensile test and the range of contraction is determined in a temperature test.

During the tensile test

the cable elongation ε_K,
the fiber elongation ε_F and
the change in attenuation $\Delta\alpha$

are registered as a function of the tensile force F. A tensile test equipment is used for this purpose that has a cable take-up similar to a tackle block (Figure 9.29) for a length of about 100 m.

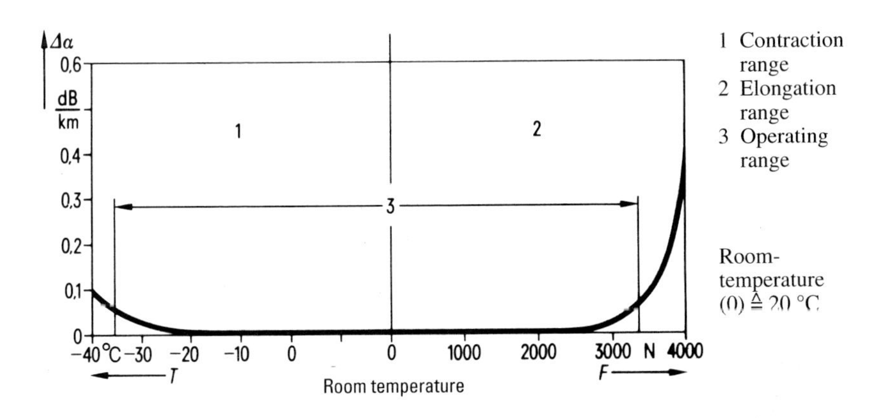

Figure 9.28
Operating window of a single-fiber and multifiber loose buffer cable (example)

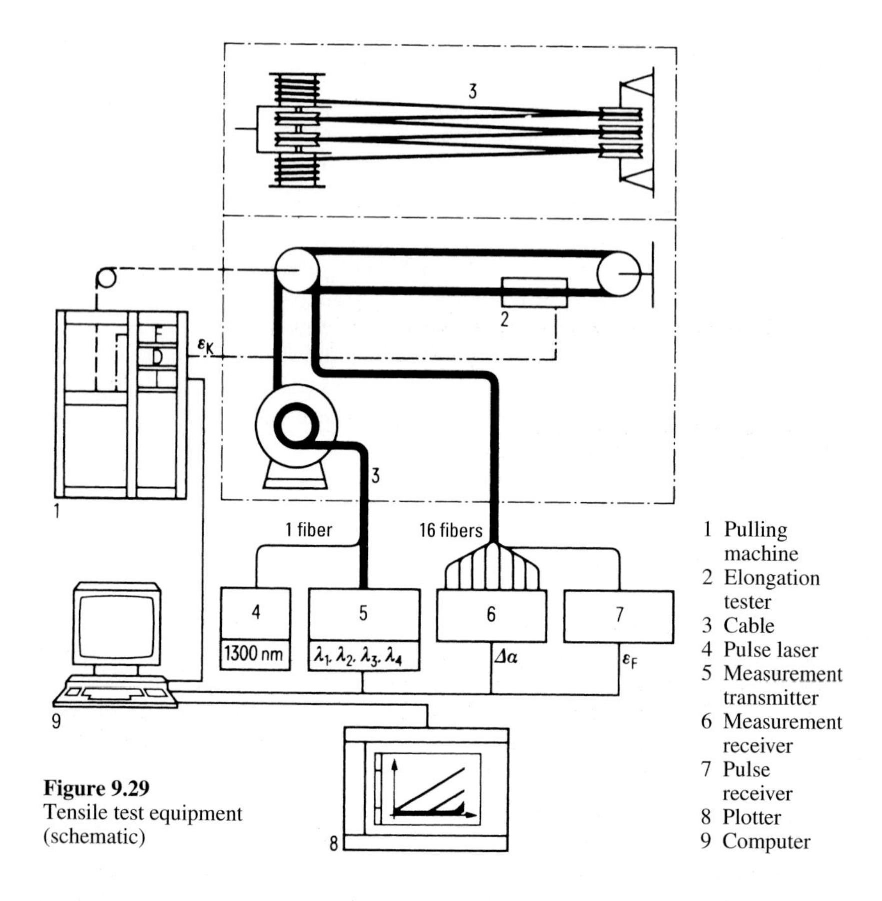

Figure 9.29
Tensile test equipment (schematic)

The cable elongation is measured by an elongation tester hooked to the outside of the cable sheath and the fiber elongation is determined by a pulse delay time measurement. The change in attenuation is measured in a special attenuation change measuring setup that makes it possible to measure several single-mode or multimode fibers at the same time and at up to four wavelengths. The corresponding data can be recorded on a tensile test record (Figure 9.30).

Figure 9.30 shows that when the fiber begins to elongate at about 3000 N the attenuation also increases. These changes in attenuation are reversible, i.e. they decrease as the cable is relieved.

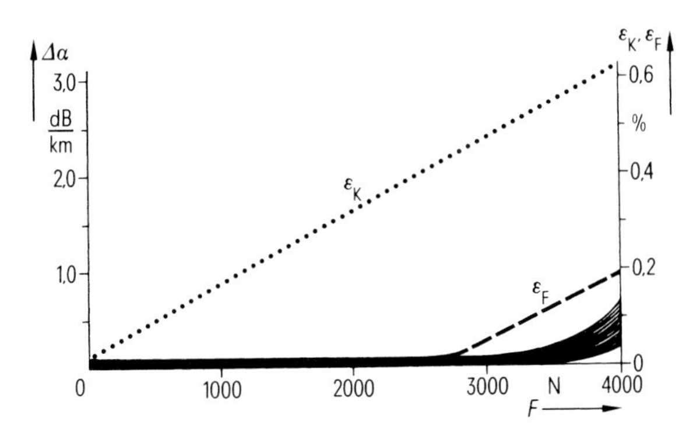

Figure 9.30
Tensile test record of a single-fiber or multifiber loose buffer cable;
$\Delta\alpha$, ε_K, ε_F as functions of the tensile force

9.6.2 Test of the Temperature Performance

Optical cables are subject to a wide variety of temperatures during operation and generally during storage, too. For buried and duct cables only relatively small variations in environmental temperatures are to be expected; by contrast for aerial cables they would be greater. For the specified temperature ranges for each cable type the transmission properties of the optical fiber may only change within a limited range.

In order to determine the temperature performance, the change in attenuation is measured as a function of the temperature. For this purpose, the cable and the drum it is wound on are placed in a computer-controlled temperature chamber with both ends attached to an attenuation measuring test set. Several software programs make it possible to run a variety of temperature cycles to meet the specified requirements. Figure 9.31, as an example, shows the average attenuation change of a multifiber loose buffer cable with eight buffers each containing six single-mode fibers.

From Figure 9.31 it can be seen that the average attenuation change at a wavelength of 1300 nm at – 40 °C is less than 0.1 dB/km.

In summary, regarding tensile and temperature tests, it can be stated that tensile and temperature curves can be plotted together on one graph to give a single description showing the operating window (Figure 9.28) of an optical cable.

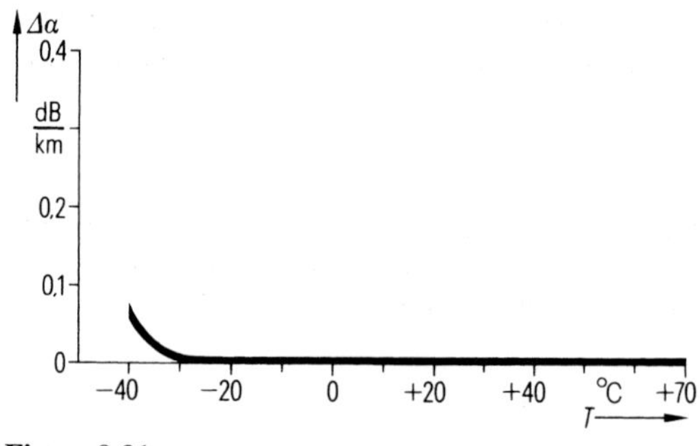

Figure 9.31
Attenuation change $\Delta\alpha$ at a wavelength of 1300 nm as a function of the temperature for a cable with eight-layer stranded multifiber loose buffers and six single-mode fibers each

9.6.3 Further Environmental Tests

In the following a list of further environmental tests is provided.

Mechanical Tests

Corresponding to the application, an optical cable may be subject to different stresses. Vibrations can occur, for example, in aerial cables or when optical cables are installed along public transportation lines, such as train tracks, or on bridges. When internal wiring is necessary for units which require movements, as, for example, in medical X-ray machines, then special demands are set with regard to cyclic and twist bend resistance.

Last but not least, to a certain degree cables must withstand lateral crunch and impact forces, which, for example, can occur during installation.

To determine these cable properties, the following tests can be carried out according to IEC standards: bending, torsion, impact, crush and vibration.

Thermal Tests

In addition to the temperature cycle tests, in order to be able to evaluate the durability of cables, aging tests are carried out at higher temperatures, e.g. approximately 70 °C over a long period of time (half a year). In this time, both transmission properties and potential changes in the individual materials are monitored.

For indoor and special cables for industry, higher flame resistance is often required, which can also be tested together with the other environmental tests.

Chemical Tests

In actual use, cables are subject to a great variety of chemical environmental conditions, e.g. to oil, alkaline solutions, acids. Corresponding tests are carried out on short lengths of the cable and any changes are monitored.

Other Tests

If necessary, special tests for rodent resistance, bacteria, microbes, termites, gas blocking must be carried out.

9.6.4 Fire Resistance

Particularly for indoor cables fire resistance is required. Several different requirements on fire restriction exist. Some of them are:

FR flame retardant
The cable is made up of special materials, such that it is difficult to ignite it by fire. For flame retardency cable construction is important as well.

NC noncorrosive
The cable sheath and the buffers do not contain materials that are corrosive, i.e. attack electronic equipment and the respiratory system. The most common corrosive materials are compounds of chlorine, bromine and fluorine. As an example, PVC produces corrosive chlorine if it burns.

LS low smoke
The materials for the cable sheath and the buffers produce only very little smoke in case of fire. This is particular important for emergency exits and escape routes.

0H halogen free
Halogen free cables are a subgroup of non corrosive cables and do not contain corrosive halogens. However, they could contain other corrosive materials. Refer also to chapter 9.2.

Commonly used combinations of requirements for fire resistance are FR NC and LS 0H.

10 Cable Plant Design

In this chapter the most important facets of planning optical cable plants are summarized. In addition to insights into installation practices for optical cables, particular attention is paid to methods of calculating transmission parameters, such as attenuation and dispersion, as well as to mechanical parameters and to the jointing technology.

An optical cable plant consists of the installed cable lengths joined to each other and extends to the first nonpermanent joints, the connectors, at their ends.

The following remarks have been kept as general as possible, in order to promote a better understanding of this sometimes complex context. For more detailed information on this subject reference is made to corresponding specialized technical literature.

10.1 Planning the Transmission Properties

Attenuation and bandwidth of the optical fibers used and the loss of the splice connections are the most important parameters affecting transmission that must be taken into consideration in planning optical cable plants. Any losses due to connectors, branching points, couplers, etc., and extra margins which must be planned for in the equipment have not been taken into account in the following considerations.

10.1.1 Planning the Attenuation for Single-Mode and Multimode Fibers

The attenuation a_K of a cable plant is calculated from the cable length L with the attenuation coefficient α_F and the number n of the splice losses a_{sp}. It holds true that

$$a_K = L \cdot \alpha_F + n \cdot a_{sp}$$

a_K attenuation of the cable plant in dB
L cable length in km
α_F attenuation coefficient in $\frac{\text{dB}}{\text{km}}$
n number of splices
a_{sp} splice loss in dB.

Because cable plants are designed for a longer period of operation, during planning extra margins must be added in to compensate for repair splices. This can be particularly important when new pieces of cable must be installed to make repairs of damage due to construction and excavation or because of rerouting of the existing cable. The extra margin of attenuation α_{Res} reserved for this purpose depends on local conditions and the importance of the plant. This margin can lie between 0.1 and 0.4 dB/km depending on the decision of the network operator.

From this attenuation a_{R} for the repeater spacing can be calculated:

$$a_{\text{R}} = a_{\text{K}} + \alpha_{\text{Res}} \cdot L$$

where α_{Res} is the extra margin of attenuation in decibels per kilometer. By using longest possible delivery lengths, every attempt must be made to achieve as little attenuation increase due to splices as possible.

Example

A route is to be planned for a 140 Mbit system with laser diodes so that it has a maximum attenuation for the repeater spacing of 30 dB at an operating wavelength $\lambda = 1300$ nm. The route length is to be 50 km, whereby cable lengths of 2000 m are to be employed, that is 24 splices will be needed with an average loss of 0.1 dB. The extra margin is set at 0.1 dB/km based on the given conditions along the route.

It is necessary to determine what maximum attenuation coefficient the optical fibers to be selected may have and whether the route can be serviced without repeaters (the aspects regarding dispersion are dealt with in section 10.1.3).

Solution

It follows from the above-mentioned equations that

$$a_{\text{R}} = L \cdot \alpha_{\text{F}} + n \cdot a_{\text{sp}} + L \cdot \alpha_{\text{Res}};$$

$$\alpha_{\text{F}} = \frac{a_{\text{R}} - n \cdot a_{\text{sp}} - L \cdot \alpha_{\text{Res}}}{L};$$

$$\alpha_{\text{F}} = \frac{30 \text{ dB} - (24 \cdot 0.1 \text{ dB}) - (50 \text{ km} \cdot 0.1 \text{ dB/km})}{50 \text{ km}};$$

$$\alpha_{\text{F}} = 0.45 \text{ dB/km}.$$

Table 10.1 Overall planning of attenuation for optical transmission systems

Transmission bit rate	Mbit/s	140			565		2488	
Fiber type		Multimode fiber (Core-D. 50 µm)	Single-mode fiber 1300 nm	Single-mode fiber 1550 nm	Single-mode fiber 1300 nm	Single-mode fiber 1550 nm	Sigle-mode fiber 1300 nm	Sigle-mode fiber 1550 nm
Optical transmitter								
Laser diode		FPR[1]	FPR	DFB[2]	FPR	DFB	DFB	DFB
Wavelength	nm	1270–1320	1285–1330	1500–1570	1290–1330	1525–1575	1260–1360	1480–1580
Spectral width	nm	<5	<5	<1	4	0.3	<1	0.25
Transmitted power (behind connector)	dBm	–5	–5	–5	–6	–8	–5	–4
Optical receiver Received power	dBm	–38	–38	–40	–37	–38	–18	–27
Maximum optical attenuation of repeater spacing incl. 3 dB system margin	dB	30	30	32	28	27	10	20
Attenuation coefficient of fiber	dB/km	1.0	0.4	0.25	0.4	0.25	0.4	0.25
Average splice loss	dB/splice	0.2–0.1	0.1–0.05	0.1–0.05	0.1–0.05	0.1–0.05	0.1–0.05	0.1–0.05
Installation length	km	1–2	1–2	1–2	1–2	1–2	1–2	1–2
Splice loss per length	dB/km	0.2–0.05	0.1–0.025	0.1–0.025	0.1–0.025	0.1–0.025	0.1–0.025	0.1–0.025
Repair margin	dB/km	0.35–0	0.2–0	0.2–0	0.2–0	0.2–0	0,2–0	0.2–0
Attenuation coefficient of cable plant	dB/km	1.55–1.05	0.7–0.425	0.55–0.275	0.7–0.425	0.55–0.275	0.7–0.425	0.55–0.275
Length of repeater spacing	km	19.3–28.5	42.8–70.5	58.1–116.3	40–65.8	49–98.1	14.2–23.5	36.3–72.7

[1]) FPR = Fabry-Perot-Resonator; [2]) DFB = Distributed Feedback

This means that the required optical fibers must have an attenuation coefficient of not more than 0.45 dB/km. No repeater is necessary.

Table 10.1 shows the planning of various transmission systems with all attenuation parameters which must be taken into consideration.

10.1.2 Bandwidth of Graded Index Fibers in Cable Plants

The bandwidth of an optical fiber with graded index profile is mainly limited by the modal and/or material dispersion. When light emitting diodes with a broader spectral width and an operating wavelength of $\lambda = 850$ nm are used, material dispersion plays the greater role. On the other hand, with laser diodes, which typically have a smaller spectral width and an operating wavelength of $\lambda = 1300$ nm, the modal dispersion predominates.

There are several methods of calculating the necessary bandwidth length product

$$b_1 = B_1 \cdot L_1$$

of a graded index fiber given a system bandwidth B at the length L of the fiber, which approximates the length dependence of the bandwidth. One method is based on the power law (Section 5.3):

$$\left(\frac{B}{B_1}\right) = \left(\frac{L}{L_1}\right)^{-\gamma}$$

B system bandwidth in MHz
b_1 bandwidth length product in MHz · km
B_1 bandwidth of the optical fiber in MHz at L_1
L length of the optical fiber in km
L_1 length of the optical fiber – typically 1 km – at the bandwidth B_1

Because, due to modal dispersion, the bandwidth does not decrease linearly with the length when several cable lengths are concatenated, an attempt is made to approximate the true length dependence by means of the so called γ factor. The values for γ generally lie between 0.6 and 1.0 so that as a general rule 0.8 can be expected.

Example

The bandwidth length product b_1 required of a graded index fiber is to be calculated for a 34 Mbit system with a repeater spacing of $L = 25$ km at an operating wavelength of $\lambda = 1300$ nm. The system bandwidth B of the 34 Mbit system is ≥ 50 MHz.

Solution

$$b_1 = B_1 \cdot L_1 = \frac{B \cdot L_1}{\left(\frac{L}{L_1}\right)^{-\gamma}} = B \cdot L_1 \cdot \left(\frac{L}{L_1}\right)^{\gamma}$$

$$b_1 = 50 \text{ MHz} \cdot 1 \text{ km} \cdot 25^{0.8} \approx 657 \text{ MHz} \cdot \text{km}.$$

Because the grouping of the bandwidth-length product of a graded index fiber at an operating wavelength of λ = 1300 nm is carried out in 200 MHz · km steps (600–800–1000–1200 MHz · km), in the above-mentioned example, with 657 MHz · km, a fiber is needed with a bandwidth-length product of 800 MHz · km.

For comparison, Table 10.2 gives bandwidth values for graded index optical fibers of digital systems at λ = 1300 nm.

10.1.3 Dispersion of Single-Mode Fibers in Cable Plants

During planning of cable plants with digital systems of up to 140 Mbit/s in which lasers will be used exclusively, the bandwidth of the single-mode fibers generally can be neglected, because it reaches far into the gigahertz range and therefore the attenuation values (fiber, splice and margins) limit the repeater spacing.

For single-mode fibers, the dispersion is given instead of the bandwidth. The calculation of the pulse broadening from the dispersion is particularly easy to carry out:

$$\Delta T = M(\lambda) \cdot \Delta\lambda \cdot L$$

ΔT pulse broadening in ps

$M(\lambda)$ chromatic dispersion in $\frac{\text{ps}}{\text{nm} \cdot \text{km}}$

$\Delta\lambda$ spectral width (FWHM) of the transmitter in nm

L length of the fiber in km

System	Bandwidth per repeater spacing	
	LED (MHz)	LD (MHz)
8 Mbit/s	25	25
34 Mbit/s	50	50
140 Mbit/s	170	120

Table 10.2 Bandwidth for graded index optical fibers of digital systems at λ = 1300 nm

Example

This is a calculation of the pulse broadening and bandwidth which result in a transmission route of the length L = 50 km when a laser with an operating wavelength λ = 1300 nm and the spectral width $\Delta\lambda$ = 2,5 nm and a dispersion $M(\lambda)$ = 3,5 ps/min · km is given.

Solution

$$\Delta T = 3.5 \frac{\text{ps}}{\text{nm} \cdot \text{km}} \cdot 2.5\ \text{nm} \cdot 50\ \text{km};$$

$$\Delta T = 437.5\ \text{ps};$$

The bandwidth of an optical fiber is calculated as explained in Section 5.4:

$$B \approx \frac{0.441}{437.5\,\text{ps}} \approx 1\ \text{GHz}.$$

In order to calculate the dispersion in systems with rates of more than 565 Mbit/s, the specific characteristics of the laser, such as mode partition noise, must be taken into consideration in addition to the above-mentioned parameters.

In planning a cable plant with single-mode optical fibers, the size of the mode field diameter is also interesting because it may affect the splice loss and the bending behavior of the optical fiber and thereby the power budget and hence the length of the route.

10.2 Mechanical Planning Aspects

The main goal of planning with respect to mechanical aspects is to construct the optical cable so that it is optimally suited to the environmental conditions and thereby provides sufficient protection for the optical fibers from external influences. For this purpose data must be collected from many sources in order to determine whether standard cable designs or special constructions should be used.

Corresponding to their applications, a differentiation is made between outdoor, indoor and special cable plants.

It is important to check several parameters, particularly with regard to conditions of the route and the installation practice. These parameters are:

Course of the Route

Maps of the route and contour maps must be used to determine gradients of ascents and descents as well as the rivers and roads to be crossed over or under, and furthermore the number of critical curves (with a notation of angles).

Nature of the Terrain

For this purpose it is necessary to know, for example, whether the terrain in question is plain, mountains, subsidence, forest, swamp, lake, etc.

Type of Soil, Type of Climate

Is the soil humus, clay or sand? Does it contain chemical contaminations? What are the average temperatures?

Installation Practice

A differentiation is made between burying and ploughing in the cables. The bending radius and the depth in which the cables are to be installed must be taken into consideration.

When cables are to be hauled in ducts, the length of the cable and the maximum possible tensile force applied to the pulling eye on the cable or to the cable grip must be considered.

Optical cables do not require a special installation practice as compared to cables with metallic conductors because they are lightweight with high flexibility and relatively small diameters.

During installation the bending radii must not be smaller than those listed in the specification of each cable. The most optimal of the installation methods listed must be chosen after consideration of the parameters listed above.

The cable construction must be designed so that the right choice and dimensioning of the structural elements prevent mechanical, thermal and chemical influences from having a permanent effect on the transmission properties. In any case, it is important that the longest possible lengths are employed, so that the amount of work and increases in attenuation due to splices are minimized. It has already become general practice to install 2000 m lengths or longer.

In a great many cases, specifically when it is possible to assure that the maximum tensile force is not exceeded, optical cables can be installed *by hand* due to their low weight. In these cases, however, it is not possible to document the tensile force applied during installation.

To determine the potential tensile force during installation, it can be assumed that in straight, horizontal cable routes the tensile forces increase linearly with the length of the cable being installed. In bends it is important to take into consideration the fact that the bending angle and the friction cause an exponential increase in the tensile force.

A further method of installing optical cables is to use pressurized air to inject them into ducts. This is possible in suitable terrain with an optimum ratio of duct to optical fiber diameter and up to a length of approximately 1000 m.

When very long optical cable lengths are being installed (for example > 3 km in one direction), one or more self-powered intermediate take-up units can be employed.

In order to better utilize ducts already installed, it is possible to pull up to four plastic subducts into them. This type of subdivision makes it possible to install several cables independently of each other.

In exceptional cases, cables with small outside diameters can be installed one after the other in ducts without subdivisions. In these cases, the maximum difference in the outer diameter between the cables installed in this manner should not be larger than 5 mm because they might become wedged or jammed.

With indoor cables inflammability and halogen content are factors to be taken into consideration. Mechanical stress on the other hand is a less important factor.

For special cables, e.g. self-supporting aerial cables, mine cables, submarine cables, etc., specific criteria must be considered in addition to those already mentioned. They cannot be dealt with here due to their complexity and number.

Power supply companies can apply optical cables with great advantage. Some possible uses are:

▷ buried and duct cables as standard optical fiber cables installed according to the above methods;

▷ self-supporting aerial cables following the techniques used with ground wires. There are cable designs with one or more layers of wire armoring made from galvanized steel, Aldrey (aluminum alloy, E-AlMgSi) or Stalum (aluminum clad steel, ACS) or a combination of them as well as fully dielectric designs. For both types even pole distances of more than 500 m and severe environmental conditions (wind and ice loads) create no problems in cables which have been dimensioned accordingly.

Installation techniques and equipment correspond to that generally employed in overhead constructions.

10.3 Fiber Optic Jointing Technique

Since cables are usually produced in standard lengths and cable networks require branching, it follows that fiber optic cables will require jointing. Consequently, cable plant designers have to take into account not only the attenuation coefficients of the fiber but also the losses introduced at the joints. In order to minimize the overall cost of the cable plant, there is always a need to maximize the distance spanned without signal repeaters. Hence, it is necessary not only to increase cable lengths and reduce inherent losses in the fiber, but also to optimize the losses at the joints.

The joints themselves may be divided into mechanical and fusion joints. The mechanical joints are further differentiated into mechanical splices (Chapter 10.3.1) and connectors (Chapter 10.3.4).

Connector joints are *nonpermanent* and comprise two fiber optic connectors held together by an adapter to ensure mechanical contact. The critical losses in connectors are the insertion loss and the return loss (Chapter 5.11). These losses are determined directly by the type and quality of the end face of the connectors.

Mechanical splices are employed mostly in networks with multimode fibers or for short distances. They are also frequently used for making *temporary* connections when carrying out repairs or measurements.

It is often the case that the requirements of high-performance communication networks cannot be met with mechanical splices, particularly when single-mode fibers are employed. In such cases, fusion splicers for optical fibers are used to establish a bonded joint by fusing the fiber ends. The fusion splices (Chapter 10.3.2) are referred to as *permanent* joints.

10.3.1 Mechanical Splices

Whereas during fusion splicing the ends of optical fibers are bonded together, with mechanical splices, as the name indicates, they are mechanically attached. This is usually achieved by means of V-groove mechanisms, into which the fibers are placed and clamped tight. At the location, where the fibers meet, a factory prefilled fluid usually equalizes the difference in refractive index between glass and air. This fluid exhibits almost identical light transmission properties compared with the optical fibers and it serves to bridge the gap between the fiber ends, as they only lie close to each other.

Mechanical splices show approximately the same low splice loss values as fusion splices, however, their return loss depends on temperature. An insertion loss of 0.2 dB is a realistic value for a field installation. Due to change in the index of refraction, especially at low temperatures (– 40 °C), the return loss is limited by conventional methods to approximately –50 dB.

Where performance is not as critical, e.g. for temporary joints, for quick fiber optic jumpers without extensive installation and for repairs, mechanical splices are extremely suitable and should be used. Therefore they provide an ideal alternative to the fusion splicing method.

Modern mechanical splices, such as the CamSplice, are simple, service-friendly and suitable for both single-mode and multimode fibers. The characteristic feature of the CamSplice is the cam mechanism, which allows the inserted fibers to be fixed without any adhesive. Together with a precise glass groove made up of two glass capillaries, this mechanism comprises a patented fiber positioning method, which guarantees an extremely precise alignment of the fibers. The average splice loss is 0.15 dB. The CamSplice can be used universally for all optical fibers with coatings of 250/250 µm, 250/900 µm and 900/900 µm. It is reusable and can be detached on one side.

For the preparation of the fibers, only a stripping tool for the removal of the coatings and a fiber cleaver is required. The prepared fiber ends are inserted into the CamSplice, until the endfaces touch each other. Then the cam lock parts are turned to fix the fibers (Figure 10.1). This splicing process does not need any adhesive or polishing of the fibers.

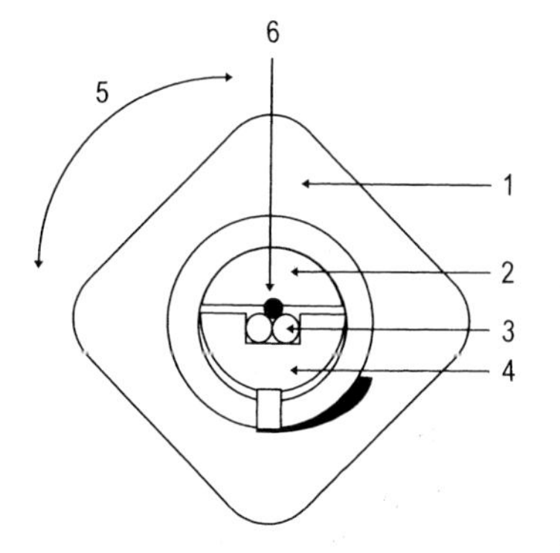

1 Cam mechanism
2 Upper part
3 Glass capillary
4 Lower part
5 Direction to turn the cam for locking and unlocking
6 Optical fiber

Figure 10.1
Functional principle of the mechanical splice, CamSplice

10.3.2 Fusion Splices

The aim of fusion splicing is to join the light-guiding cores of the fiber in a way that provides mechanical strength, long-term stability together with low loss and low reflection. This is achieved by directly fusing the fibers using a fusion splicer.

Splicers for optical fibers are used for precisely aligning the fibers in relation to each other and for fusing the ends of the fiber together by means of an electric arc. This arc is generated by means of a high-frequency alternating voltage between two electrodes (Figure 10.2).

Unlike mechanical fiber splicing, fusion splicing has a self-centering effect on the fiber thereby automatically optimizing the splice loss. The self-centering effect is the tendency of the fiber to form a homogeneous joint which is consequently free of misalignment as a result of the surface tension of the molten glass during the fusion bonding process. Due to that it is possible for fusion splicers to achieve mean losses of less than 0.1 dB without core-to-core alignment.

Fiber optic splicers can be divided into units for single-fiber and fiber ribbon applications. Owing to the different dimensions and fiber counts (Figure 10.3) there is a need for splicers which are suitably equipped for the particular applications.

Single-fiber fusion splicers are available both with and without core-to-core alignment. Units with core-to-core alignment provide splice losses of less than 0.05 dB for single-mode fibers.

The fusion splicing of multimode fibers is not as critical owing to the large core diameter and large numerical aperture. Splice losses of less than 0.02 dB are typically obtained.

Splicers for fiber ribbons exploit the self-centering effect. Core-to-core alignment of fiber ribbons is feasible owing to the fixed spacing of the fibers relative to each other. Mean losses of less than 0.1 dB with maximum losses of 0.15 dB can be achieved per splice depending on the quality and quantity of fibers in the ribbon. If these losses are acceptable in the application, all the fibers in the ribbon can be prepared and spliced simultaneously thus saving time and increasing productivity.

Fiber ribbon splicers often are used for single fibers.

The use of single-fiber and fiber ribbon cables differs from region to region and is affected by special network structures. Commonly used fiber ribbons comprise 4, 8 or 12 individual fibers, although no standard has yet been

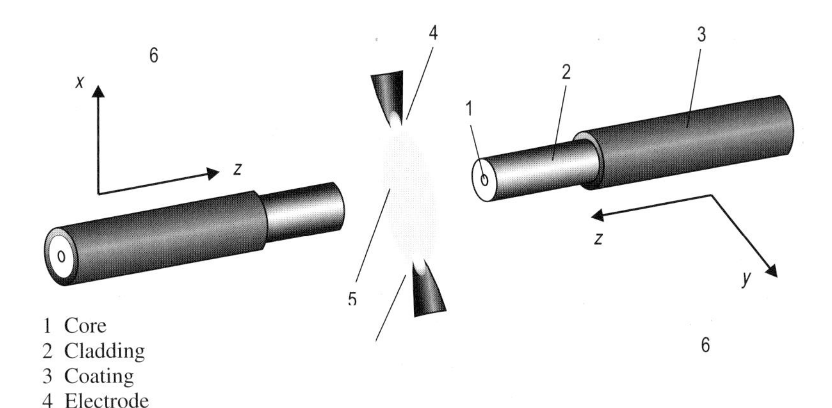

1 Core
2 Cladding
3 Coating
4 Electrode
5 Electric arc
6 Fiber alignment

Figure 10.2
Principle of fusion splicing

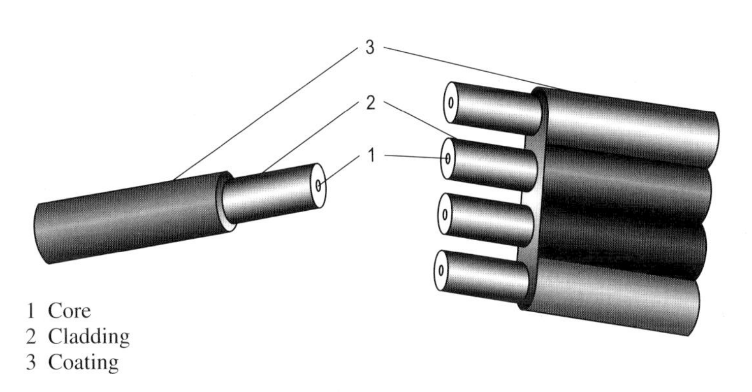

1 Core
2 Cladding
3 Coating

Figure 10.3
Single fiber and fiber ribbon

approved. They are currently employed in four- and eight-fiber ribbon networks in Asia and parts of Europe, as well as in twelve-fiber ribbon networks in the USA and Asia.

Fiber preparation for ribbons is more complex than for single-fiber applications, e.g. due to simultaneous stripping and cleaving of all fibers in the ribbon structure. It is also necessary to simplify the handling of different ribbon sizes, therefore splicers employ handling adapters in ribbon applications.

Figure 10.4 Cleaver A8

The splicing process is basically the same for single fibers and fiber ribbons. It can be split up into three steps:

Fiber preparation
Fusion splicing
Protection and storage of the splice.

When splicing fibers, it is always important to ensure that each process step is performed cleanly and accurately.

Fiber Preparation

Regardless of the jointing technique, the fiber preparation comprises the following process steps:

Stripping

After the workplace and cable have been prepared, the buffer and coating around the fiber must be removed. For single fibers manually operated tools are usually used for this purpose. Fiber ribbons are normally stripped using thermal stripping tools. These tools apply heat and a controlled cut to the ribbon coating to strip the coating from all the individual fibers in the ribbon simultaneously.

Figure 10.5 Cleaver D6

The stripping tools must be designed and operated to ensure that no damage occurs to the fiber. Any damage to the fiber could produce microcracks causing the fiber to break during handling or splicing. The tensile tester incorporated in most splicers is used to check the unprotected splice and verify its mechanical strength on storage in the closure or splice module.

Cleaning

The exposed fiber ends are cleaned with alcohol to remove coating residues and other contaminants.

Cleaving

The fibers are cut using a cleaver. The cleaved face must be as flat and perpendicular to the fiber axis as possible because the quality of the end face directly influences the splice loss. High quality cleavers typically achieve cleave angle deviations of 0.5° to the perpendicular. There are different types of cleavers available to suit the various applications and requirements. Figures 10.4 and 10.5 show typical high performance cleavers for cleaving single fibers. They can be used for all common types of fibers for various cleave lengths and are easy to operate.

In order to minimize the risk of damage in transit or operation under field conditions, there is often a provision on fusion splicers for mounting the cleavers on the work area of the splicer case.

Fusion Splicing

In principle, fusion splicing is the same on all fusion splicers for optical fibers and consists of the following steps:

Insertion of the fiber ends

Insertion of the prepared fiber ends in the splicer's fiber holders. This may involve coarse alignment of the fiber ends by manual adjustment.

Fiber alignment

This step is performed fully automatically in most high quality splicers. A distinction is made here between various types of fiber alignment and various forms of fusion process control (Chapter 10.3.3). On simple splicers the fibers are aligned by inserting and adjusting them manually with acceptable accuracy.

Arc cleaning

Many splicers permit additional cleaning of the fiber ends by brief arc ignition to burn off or blow away dirt particles or residues in the splicing area.

Fusion with fine adjustment while arc ignited

The fusion process is generally performed by the splicer automatically according to preset parameters. On splicers with the LID System™ and AFC™, it is additionally controlled by direct or indirect measurement of the splice loss (Chapter 10.3.). The process is often subdivided into preliminary and main fusion phases. In the preliminary fusion phase, the fiber ends are melted to ensure optimum bonding of the fibers during the main fusion phase.

Splice loss evaluation or measurement

High quality splicers provide this capability, so that the splice loss can be assessed directly after the fusion process.

Tensile testing of the finished splice

Many splicers provide an integrated tensile tester. This is used to check the mechanical strength of the completed splice.

Protection and storage of the splice

Protecting the splice

The splice is removed and a splice protector (e.g. heatshrink or crimp splice protector) is usually applied to the bare glass fiber in a facility provided for this purpose. Splice protection facilities are often built-in or available as adaptable accessories.

Storing the splice

The protected splice is stored (usually in a splice organizer) in a splice tray (Figure 10.6). Splicers can be equipped with splice tray holders. When storing the protected splice and the fiber slack, it is essential to observe the maximum bending radius or the fibers.

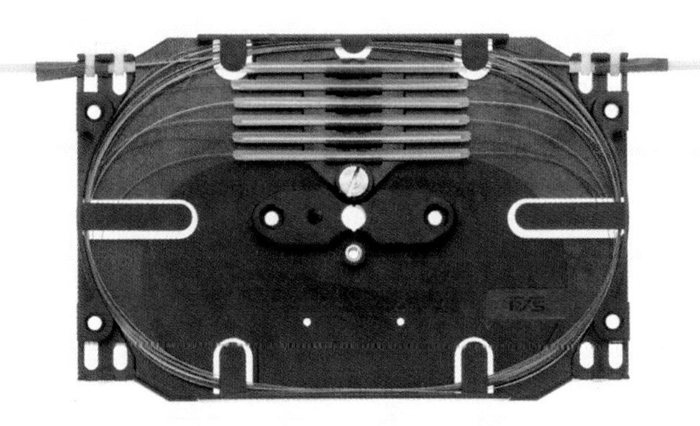

Figure 10.6 Standard splice tray with heatshrink splice protectors

10.3.3 Splicers for Optical Fibers

Modern fusion splicers offer a wide variety of options to suit every application (Figure 10.7). The units differ mainly in their features, such as:

Fiber alignment and fusion control systems
Types of fiber alignment
Physical design

Fiber alignment and fusion control systems

Splicers often employ fusion process control to permit precise fiber alignment as well as splice loss evaluation. Two different systems are used for this purpose:

▷ Systems in which the mechanical properties of the fiber are evaluated visually by video image.

▷ Systems which measure the response of light transmitted across the spliced joint.

In the following paragraphs the most frequently used alignment and fusion control systems are explained.

Profile Alignment System (PAS)

This system employs magnifying optics and a video system to determine the outer dimensions and the position of the cores in the fiber in two perspectives. Video image evaluation of the two recorded views of the fibers is used to control the fiber alignment. This system allows core-to-core alignment of the fiber ends. The fibers are displayed on a monitor. Owing to the complexity of the focusing, only one view is shown at a time.

Control of the fusion process is effected solely through the fiber alignment and preset fusion parameters. Optimization of the parameters must be performed before splicing.

The splice loss is estimated by assessing the fiber and core misalignments after the splicing process. The indicated losses are based on a comparison of the detected misalignments with values recorded in a series of tests.

Lens-Profile Alignment System (L-PAS™)

This system is a simplified form of PAS. L-PAS™ employs the lens effect of the fiber to control the alignment. Given the standard optical fibers with their good core concentricity, this system is adequate for applications with average requirements.

Figure 10.7 Splicer family

The L-PAS™ system, unlike PAS units, needs only simple and therefore robust optics. There is no need to move the CCD cameras for focusing. This makes it possible to reduce the size of the equipment while displaying both fiber views simultaneously on a built-in monitor.

The technique employs a video evaluation system that assesses the fiber end-face, detects contamination and mechanical damage, facilitates alignment and evaluates the splice loss (Figure 10.8).

The L-PAS™ system provides extremely short fusion processing times of typically 10 seconds including fiber alignment.

Local Light Injection and Detection System (LID System™)

The LID System™ is a built-in transmitted light measuring system operating in the single-mode transmission region at a wavelength of 1300 nm. The LID System™, unlike the optical evaluation systems, performs fine alignment and fusion control as a function of the transmission characteristics of the joint. This system uses a bent-fiber coupler (source) to inject light into one of the fibers to be spliced. On the opposite side of the splice the transmitted

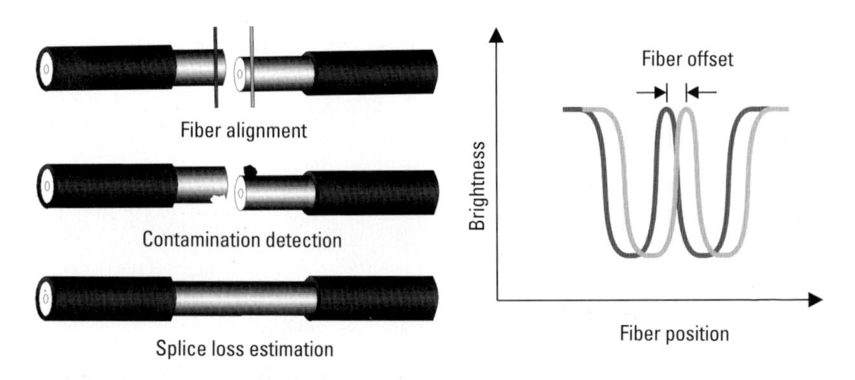

Figure 10.8 Principle of the L-PAS™

light is extracted by a bent-fiber coupler and detected by a photodiode. The electrical signal derived as a function of the LID transmission level is evaluated and used e.g. for microprocessor-controlled core-to-core alignment as well as automatic fusion time control AFC™ (Figure 10.9). The bent-fiber couplers are designed to avoid any impairment of the fiber's characteristics (fiber and coating).

The automatic fusion time control AFC™ permits "on line" optimization of each individual splicing operation to obtain the lowest possible splice loss. Environmental influences and different fiber characteristics are taken into account automatically. This optimization, like the core-to-core alignment, is

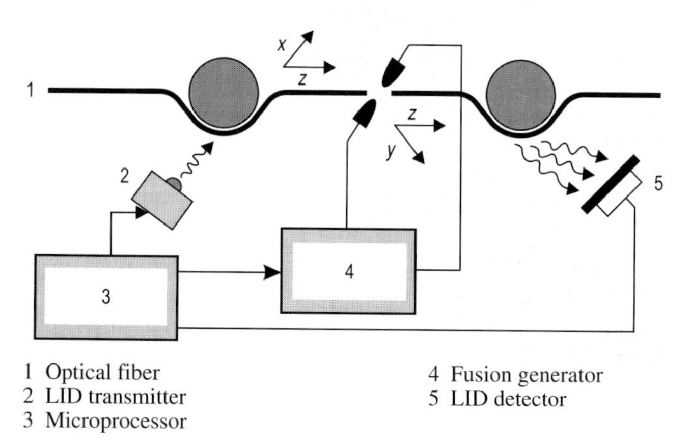

1 Optical fiber
2 LID transmitter
3 Microprocessor
4 Fusion generator
5 LID detector

Figure 10.9 Principle of the LID System™

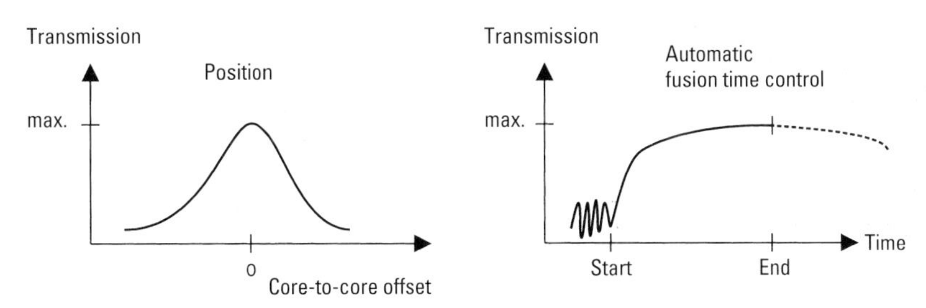

Figure 10.10
Alignment and automatic fusion time control as a function of transmission

performed as a function of the joint loss, not merely by external visual appearance (Figure 10.10). Excellent splice results are thus achieved even for non-identical fibers.

Splicers with the LID System-measuring capability (Figure 10.11) additionally perform precise cleave angle assessment and a splice loss measurement. This splice loss measurement provides reliable splice loss values because the LID System™ has an operating wavelength of 1300 nm, thus simulating the

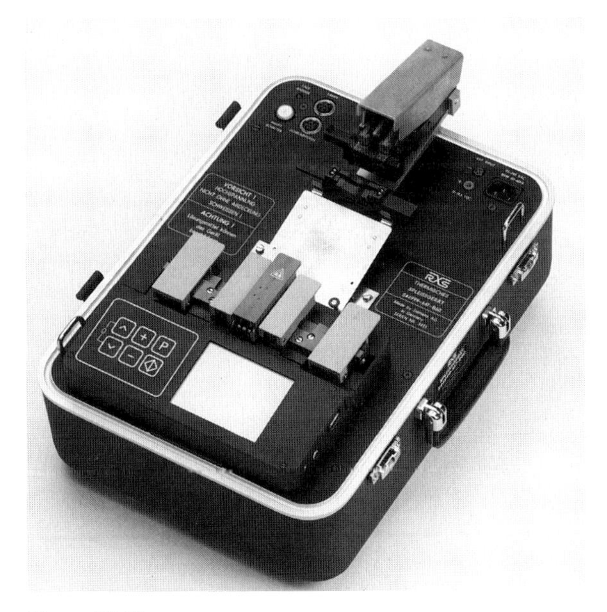

Figure 10.11
Splicer X60 with L-PAS and LID System with measuring capability

real operating conditions subsequently encountered (also valid when operating at 1550 nm). The otherwise customary analysis of the individual splice measurement is not required here. In addition the LID System™ measuring facility can be used to provide high-precision attenuation splices. These are used as attenuators for power matching in passive optical networks and can only be obtained in this quality as preconnectorized components.

Splicers with LID System™ often utilize the L-PAS™ for rapid coarse alignment and endface assessment. These units thus provide a splicing process comprising high-precision core-to-core alignment and AFC™ optimization. A splice can be completed typically in less than one minute.

Types of fiber alignment

The main aim of fiber alignment is to ensure that the fiber ends to be spliced are aligned to each other as accurately as possible. Single-mode fibers in particular demand highly accurate fiber alignment due to their small core diameter. The alignment facility is either fixed or adjustable depending on the design of the particular splicer. In core-to-core alignment the adjustment is made in the three axes x, y and z (fiber axis). On units for special applications (e.g. polarization-maintaining fibers), it is also possible to rotate the fibers.

The alignment process is performed automatically or manually. There are also combinations of these options available.

Manual alignment

Manual alignment is performed using setting wheels (e.g. micrometer screws) or by positioning fibers to stops or in V-grooves. In a splicer with fixed V-grooves or stops, the fiber ends cannot be adjusted vertically or horizontally. They are aligned in the relevant guide by their outer diameter. The alignment's sensitivity to dirt is greater on these splicers than on units with 3-axis alignment.

Automatic alignment

Automatic alignment is effected through stepping motors and/or piezoceramic alignment elements. These alignment elements consist of electronic components which respond to the application of an electric voltage by enlarging or reducing their mechanical dimensions precisely in very fine mechanical resolutions (< 0.1 µm).

Figure 10.12 Fusion splicer series X7...

The fibers are held or clamped by means of fiber guides. These differ according to the following criteria:

Material	e.g. etched silicon V-grooves, metal V-grooves, metal angles, ceramic mountings
Adjustability	fixed, movable, rotatable
Interchangeability	for adaptation to different fiber diameters or clamping methods
Accuracy	The dimensional tolerances for holding the fiber must be as small as possible in order to provide an optimum for low loss splicing.

Physical design

Splicers differ, sometimes markedly, in their construction, dimensions and weight.

Many units are available in small, compact designs and can be configured with appropriate accessories to suit specific applications in modular fashion (Figure 10.12).

Some fusion splicers form a complete workstation. Others are handy units that can be upgraded with accessories to form complete workstations. Fusion splicers with their various battery options may be operated independent from an external power supply.

High quality fusion splicers provide extensive software permitting convenient, straightforward operation via a built-in monitor and control panel. This allows the user to follow the splicing process visually, select various languages, store the splicing data and output it via an interface, freely select each individual parameter, create and store his own programs and operate in fully automatic or manual mode.

Modern fusion splicers with their intelligent software and microprocessor control permit fully automatic splicing in less than one minute with one button operation. The fusion splicers X77 and X60, for instance, achieve splice losses of typically less than 0.05 dB on single-mode fibers.

10.3.4 Connectors

The most frequently used nonpermanent fiber optic joints are fiber optic connectors. Optical fibers in a cable must be terminated most often by means of fiber optic connectors in order to make them compatible with the electronics.

Therefore fiber optic connectors represent the interface between the fiber optic cable and the active transmission equipment. In addition fiber optic connectors are employed to facilitate easy and quick changes in circuits and configurations of the fiber optic cable route. Thereby they contribute significantly to the flexibility of fiber optic cable networks.

A fiber optic connector joint usually consists of two fiber optic connectors, which are put together in a sleeve according to the butt coupling principle. It is a characteristic principle of butt coupling that the optical exit and entrance faces of the fibers are arranged in parallel and close proximity or even in

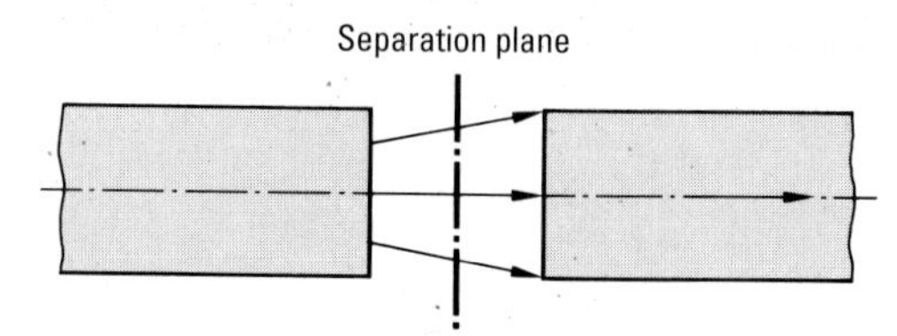

Figure 10.13 Principle of butt coupling

physical contact to one another (Figure 10.13). Only with this arrangement it is possible to produce low loss connector joints for the wavelength ranges 850, 1300 and 1550 nm.

Fiber Optic Connector Parts

The most important parts of a fiber optic connector are (Figure 10.14):

Ferrule
Housing
Crimping ring
Boot

In practical application of the above mentioned butt coupling principle, each of the optical fibers are bonded into the precise center of a cylindrical bore, a ferrule, with a very small inner diameter. The inner diameter of these ferrules ranges from 126 to 127 µm for single-mode fibers and from 127 to 128 µm for multimode fibers, when the fiber diameter is 125 µm. The end face of the ferrule is grounded and/or polished in several steps so that a clean scratch-free and extremely fine surface is created.

The ferrule which has been prepared as described above must be completed with the remaining parts (Figure 10.14) to form a fiber optic connector. In order to configure a fiber optic connector joint, two connectors are placed into a sleeve with their polished end faces butt coupled. Although there are many different types of fiber optic connectors, a de facto standard for a ferrule diameter of 2.5 mm has emerged. Connector ferrules can be made of metal, ceramic and new, durable plastic. Zirconium oxide ceramics have proven to be the material with the best over all properties for connector ferrules and are preferred over metallic materials such as silver nickel or tungsten carbide.

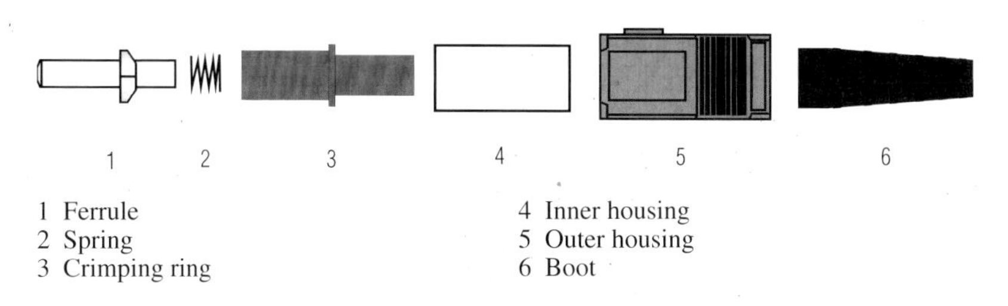

Figure 10.14 Connector parts

New, durable plastic ferrules are emerging in low cost applications, when performance is not as critical, such as in the US market for multimode fiber optic connectors. In technical performance they usually do not match ceramic ferrules, but they have a cost advantage.

Connector housings which surround the connector ferrules serve a twofold purpose. They determine the manner in which the connector is affixed to the sleeve, e.g. screw, bayonet or push-pull connection. And in conjunction with a crimping ring, the housing assures that the fiber optic cable is terminated and protected against tension.

Fiber Optic Connector Quality

The most important criteria determining the quality of single-mode fiber optic connectors are the insertion loss and the return loss. For multimode connectors usually only insertion loss is measured. The insertion loss describes the amount of increase in attenuation of a fiber optic transmission route when a connector is inserted. The magnitude of the insertion loss mainly depends on the following factors:

Axial offset of the fibers (Figure 10.15),
fiber end face separation (Figure 10.16),
angular misalignment of the fibers (Figure 10.17) and
excessive polishing of the end face, air gap (Figure 10.18).

In addition to these factors the insertion loss of fiber optic connectors also depends on the tolerances of the connector parts. Modern connectors such as

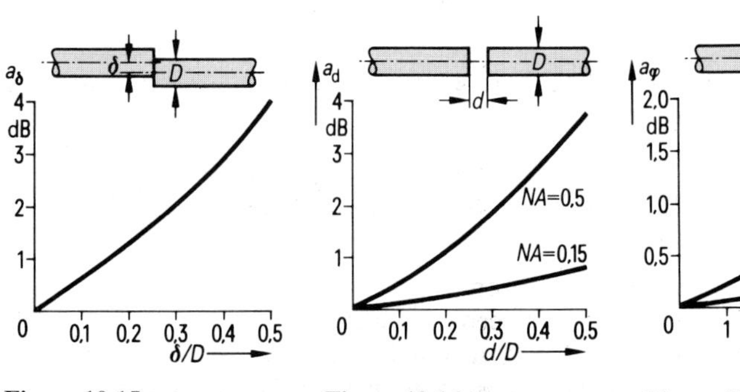

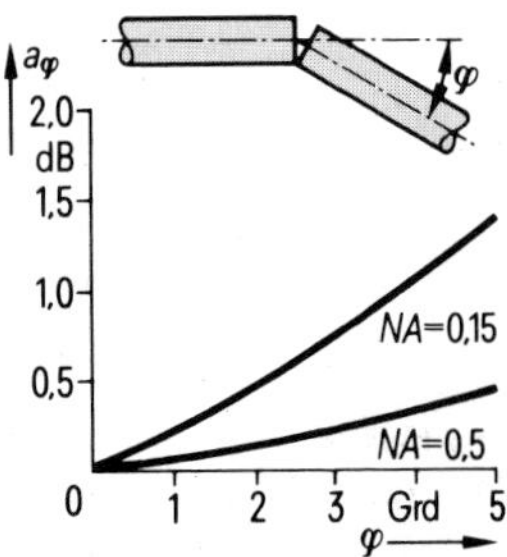

Figure 10.15
Axial offset of the fibers

Figure 10.16
Fiber end face separation

Figure 10.17
Angular misalignment of the fibers

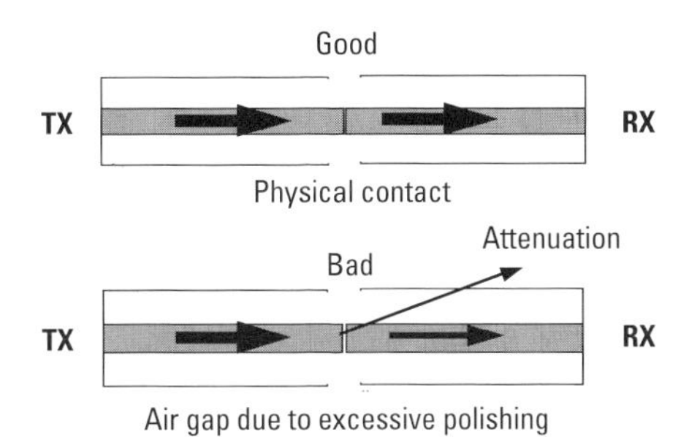

Figure 10.18
Excessive polishing of the end face

the SC or the ST connector typically show loss values of approximately 0.2 dB. In practical applications, a fiber optic connector joint, which consists of two connectors and a sleeve, should have an insertion loss of less than 0.8 dB.

Single-mode connectors should show better return loss values of less than approximately –45 dB or –55 dB. Lasers used for transmission are very sensitive to reflections. For analog signal transmission, e.g. in CATV networks, an even higher return loss of less than –55 dB is required. In practice, this is achieved by polishing the connector end faces so that an angle of 8° or 9° is created (depending on the manufacturer).

Types of Fiber Optic Connectors

There is a wide variety of fiber optic connectors available:

SC, FC, ST, FDDI, E2000, DIN, SMA, D4, EC, BICONIC, ESCON, MT.

Among these the following connectors are most widely used nationally and internationally:

SC, ST (Figure 10.19) and FDDI in LAN applications.
SC, FC, E2000 and DIN in telecommunication and CATV networks.

In general it can be observed that connectors with screw connection such as FC and DIN are slowly being replaced by connectors based on the push-pull principle. Push-pull connectors permit a higher packing density in the distribution frames and patch panels, because no space needs to be allowed for the screwing process. In addition there is no danger of overtightening the connector and scratching the polished connector, when screwing it in.

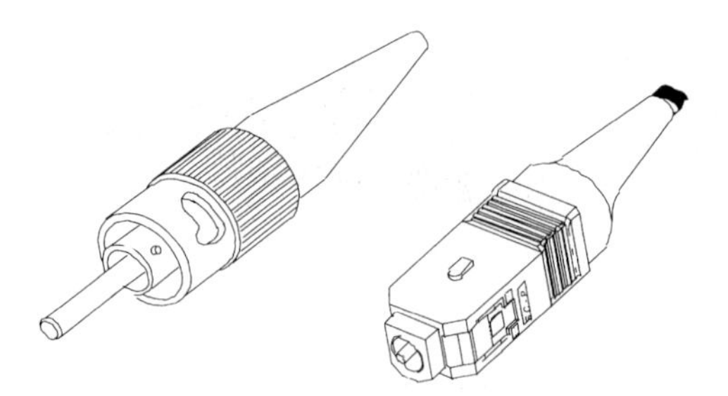

Figure 10.19 ST and SC fiber optic connector

Field Installable Fiber Optic Connectors

In LAN applications, the field installation of fiber optic connectors is often required. In this case, the installation steps usually carried out in the factory such as gluing, curing and polishing are simplified for performance in the field. It must be taken into consideration that under most field conditions the quality of connectors assembled in factories cannot be achieved. Even if factory polishing machines were used, the test equipment necessary for checking and verifying the quality of the assembled connectors could not be handled in the field. Therefore field installation based on the gluing and polishing method are normally used in multimode LANs with less demanding specifications. In applications with single-mode connectors gluing and polishing methods are not used in the field because of quality and cost.

A very practical and high quality field installable connector system is the FuseLite technique. In the factory a short length of optical fiber has been glued into the connector and polished at the connector end face (Figure 10.20). During the field installation a fusion splice is then carried out inside of the connector itself. Therefore no gluing and polishing is required. For the actual jointing a fusion splice is applied, which is recognized to be the best method to connect optical fibers. A simple method for multimode applications is the Unicam connector. A mechanical splice is used in place of the fusion splice, however, the fiber in the connector is glued and polished in the factory.

10.4 Fiber Optic Distribution Hardware

Suitable fiber optic hardware components are required at all points in any fiber optic network where fiber optic cables are terminated, branched, connected or distributed. To determine the correct fiber optic hardware for the network the following basic questions must be answered:

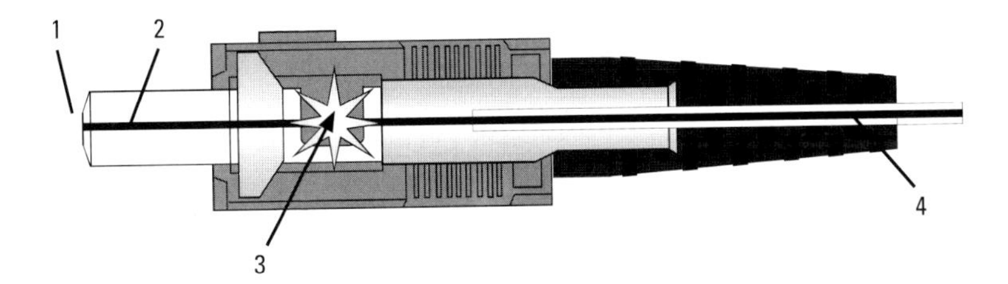

1 Factory polished connector end face
2 Factory glued in piece of optical fiber
3 Fusion point
4 Inserted optical fiber

Figure 10.20 Schematic of a FuseLite connector

Is the hardware utilized indoors or outdoors?

Indoors: Fiber optic distribution components
Outdoors: Fiber optic closures.

Shall the hardware be mounted in 19" cabinets or on the wall?

19" cabinets with fiber optic modules
Fiber optic panels according to ETSI or appropriate standard
Wall-mounted panels.

Shall field installable fiber optic connectors or preconnectorized pigtails be used for cable termination?

Field installable connectors: No splice tray required in fiber optic panel
Pigtails: Splice tray required in fiber optic panel.

What is the fiber count of the cable to be installed?

In addition to fiber optic closures for connecting and distributing cables outside of buildings, hardware components are required for use in distribution points inside of buildings or distribution centers to perform specific tasks. These distribution components shall terminate the fiber optic cable plant so that the individual cable elements, splices and connectors are easily accessible to facilitate changes in the network configuration. Since fiber optic distribution components often represent the interface with the transmission electronics, it is also important that cut-over and failure measurements can be performed without difficulties. In public or telecommunication networks the fiber optic distribution components often represent a clear demarcation of the legal areas of responsibilities of the network.

Figure 10.21
Fiber optic indoor distribution component

Planning for the points at which the fiber optic distribution components (Fig. 10.21) are placed inside of the network also depends on the total available attenuation budget. Therefore the number of separation and distribution points should be kept to a minimum.

There are four functionally different types of indoor hardware:

1. Splice modules / splice boxes / wall panels to hold, store and organize splices and couplers in trays.
2. Connector modules / patch panels / wall distribution panels to distribute, connect and jumper cables with fiber optic connectors.
3. Modules / wall panels to hold some spare length of cable or additional cables.
4. Combinations of the fore-mentioned three modules / wall distribution panels, mostly as combined splice or connector housings.

Fiber optic distribution components serve to organize incoming and outgoing fiber optic cables and determine the choice of jointing technique. They facilitate later changes in the configuration or extension of the network without requiring extensive installation work. They must also allow the application of various cable termination techniques such as fusion splicing of pigtails, mechanical splices or field installable fiber optic connectors including FuseLite connectors.

Modern fiber optic distribution components are designed as modules and are suitable for all commonly used fiber optic connector joints and couplers. The following basic requirements for fiber optic distribution components are frequently called for:

Universal application

Fiber optic distribution components must be suitable for all voice, data and video network applications.

Compatibility

All fiber optic distribution components must be compatible with each other.

For example it must be possible to combine one or more splice modules and patch panels in one distribution cabinet.

Expandability

The fiber optic distribution components must permit future extensions of the network.

Installation friendliness

Quick and simple installation must be assured. Cable termination, cable layout, space for spare cables and possible patches must therefore be clearly defined.

Depending on the application and type of the fiber optic network, different designs of fiber optic distribution components are deployed. In private networks and LAN applications the 19″ technology as known in electronics is predominant. In this case 19″ modules with different functions are inserted into the 19″ racks or cabinets at the main distribution points of the network (Figure 10.22). Depending on the size of the network either fiber optic distribution panels or combined copper/fiber optic cabinets may be used. The 19″ technology continues to be in use in many countries in Asia and in the USA for telecommunication and CATV applications. The multitude of possible applications for this technology span an entire range from installation of a single 19″ module for 12 optical fibers in a cabinet to over 1000 optical fibers in a fully utilized 19″ fiber optic distribution rack.

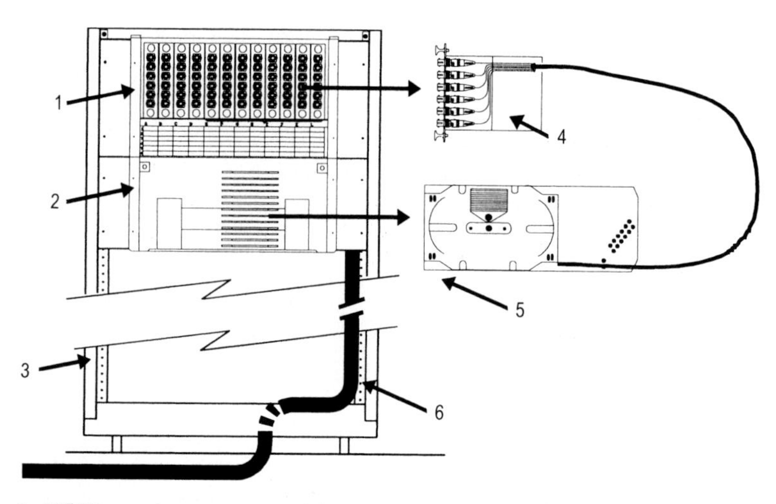

1 19″ fiber optic connector module
2 19″ fiber optic splice module
3 19″ distribution cabinet
4 Pigtail module with 6 preinstalled pigtails
5 Fiber optic splice tray
6 Incoming fiber optic outdoor or indoor cable

Figure 10.22 Schematic of a 19″ cabinet

In Europe the rack design type 7R with its slim dimensions was in use for a long period of time in telecommunication applications. Because this rack row was designed for telecommunication trunk networks with lower fiber counts, only 24 to 75 optical fibers needed to be accommodated and organized in one rack. After the fiber optic trunk networks were completed in

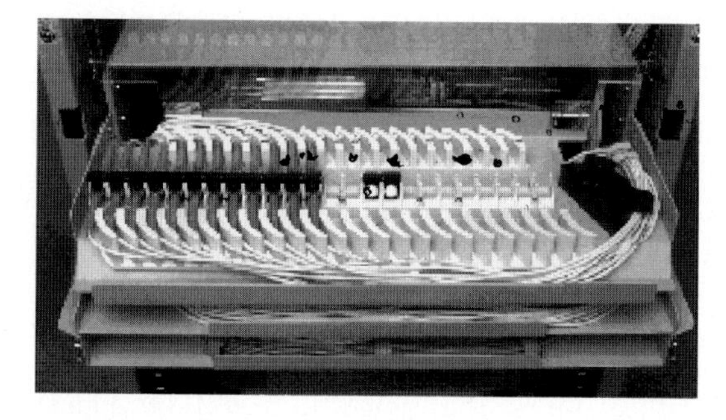

Figure 10.23 Fiber optic module according to ETSI standard

many European countries, the next step was to build fiber optic distribution networks. They require a much higher number of optical fibers. Therefore a new design type was created by the leading European telecommunication operating companies. Their standards body, ETSI, set the standard pr ETS 300 119-3, which is a metric standard in contrast to the 19" technology. The standard stipulates system racks with the dimensions (width x height x depth) of 600 mm x 2200 mm x 300 mm, which can accommodate modules in height units of 25 mm (Figure 10.23).

In small fiber optic networks, on floors in buildings and even for customer access points fiber optic wall-mounted panels are frequently used.

10.5 Closures

In general, fiber optic cable closures are employed to protect splice joints of trunk, branching and distribution cables. Additionally, they must assure the mechanical as well as the electrical continuity of the cables in such a way, that the cables function as if they were uninterrupted at this splice point. This must hold true independent of the placement of the cables, either directly buried or in cable ducts or in aerial cable routes.

The mechanical continuity of the cables is guaranteed by affixing the cable sheaths, the central members and any aramid yarns, if used. In order to protect the installer and the electronic equipment the electrical continuity is assured by electrically connecting the metallic elements of the cables. This can be achieved by connecting the armoring, the metallic sheath elements such as metallic tapes or any metallic central members with each other or with external grounding points.

Cable closures for fiber optic cables should also allow the organization of the splices in suitable splice trays and the storage of spare lengths of the buffer tubes. Of course, increased attenuation of the optical fibers, for example due to too tight bending radii, must be prevented when the cables are sealed inside of the cable entry holes, when the spare lengths of the loose buffers are stored or when the splice joints and the spare fiber lengths are placed into the splice trays.

Types of Fiber Optic Closures

The multitude of different requirements and various application characteristics cannot be met by a single closure design. Hence a complete range of closure types has been developed – the family of the Universal Closures (UC), providing solutions for all application problems (Figure 10.24).

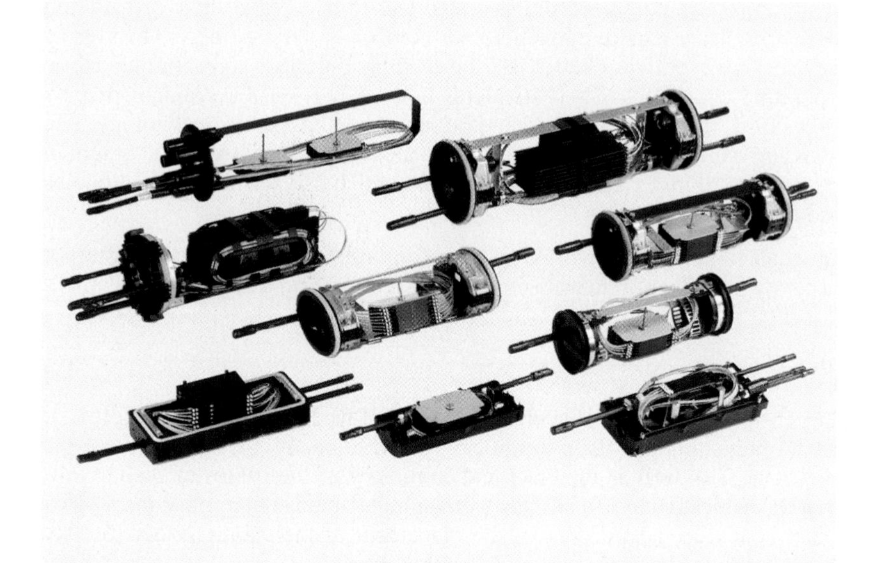

Figure 10.24 Family of Universal Closures

According to what criteria does the designer of cable networks select a cable closure? There are three aspects to be considered:

Capacity of the closure

Both the splice capacity and the number of cables to be inserted and their outer diameters must be taken into consideration.

Exterior construction of the closure

It is the shape of the closure that is important here. The following options are available:

Straight closures:

The splice is straight and the cable entry and exit holes are located on the opposite end faces of the closure tube. They are mainly used with trunk cables, e.g. between two switching offices or distribution centers, with high fiber counts in the cables (Fig. 10.25).

Figure 10.25 Universal Closure UCN 7-10

Canister closures:

This closure design has only one end face for sealing the entering and exiting cables. Canister closures are most suitable for branching cable in ring or star cable networks with medium fiber count of the cables (Fig. 10.26).

Half shell closures:

The body of the closure is composed of two half shells. This design is very compact and is mainly deployed for distribution networks with low fiber counts in the cables (Fig. 10.27).

Interior construction of the closure

In addition to adaptation to the different cable designs such as slotted core, maxibundle or loose tube cables, each in single fiber or fiber ribbon technology, adaptation to the corresponding network configuration is a further important characteristic for the interior construction of the closure. The following designs are available:

Closures for trunk networks:
Trunk networks constitute the direct connection between switching offices or between distribution centers. The trunk cables usually have a high fiber count. Post-installation branches off these cables are usually not planned, i.e.

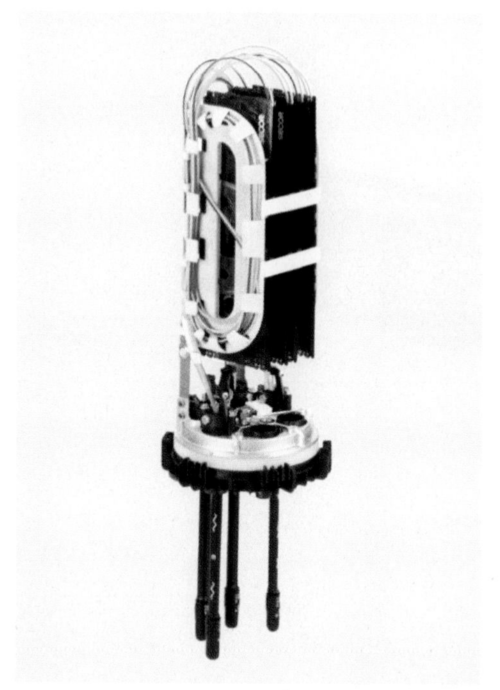

Figure 10.26
Universal Closure UCTL 6–20

future access to individual splices is not anticipated. Therefore for greater economy, the splices can be grouped in bundles and placed in splice trays. For greater packing density, usually several splice trays are grouped to form packs. Typical examples for this application are the various sizes of the Universal Closures of the type UCN (Figure 10.25).

Closures in ring or star networks:
These networks which are usually LANs with medium fiber counts are planned for today and the near future. The closures at the branching points must therefore meet two requirements.

1. The main cable must be connected with the branching cables for already existing subscribers. The splice joints with spare lengths for future jumpering must be placed in corresponding splice trays.
2. In case of branches which will be activated in the future, the main (ring) cables are often looped through the closure uncut.
   In these cases, a sufficient length of the sheaths of the cables is stripped off in order to store their loose buffers uncut with a spare length in the clo-

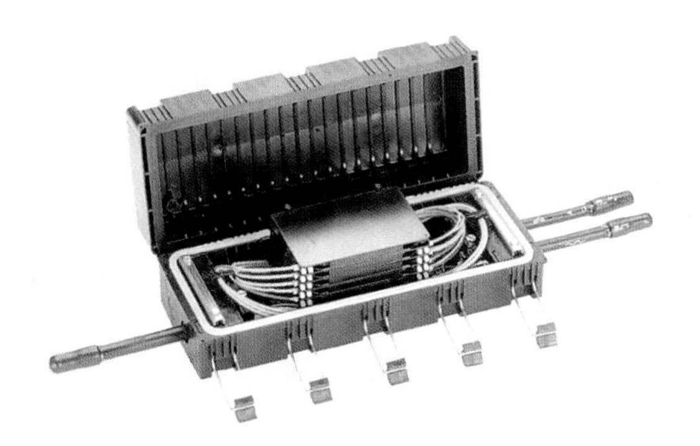

Figure 10.27
Universal Closure UCAO 4–9

sures. Only on demand, e.g. a new housing development, an industrial park or a shopping center, the required buffer tubes are cut and their fibers are joined to those of the branching cables and stored in splice trays which are now installed. The fibers are usually grouped in bundles in the splice trays. For these networks, especially the Universal Closures of the type UCTL 6–20 are suitable (Figure 10.26).

Closures in networks with sensitive data transmission:
The optical fibers of subscribers (e.g. government agencies or banks) with high bit rates and sensitive data transmission require special security protection. The fibers of these subscribers are placed in separate splice trays in closures throughout the whole cable network. Thereby it is assured, that during jumpering work only the fiber to be jumpered is moved, all other fibers remain undisturbed. In this case, a so-called single fiber management is said to be in place. The universal closure which permits such a management is the UCTL 6–20 E.

Closures in the distribution network:
Cables which are distributed to several subscribers usually contain only a small number of fibers. Typical applications for distribution closures are the distribution points of CATV networks. The splices of these closures are stored in groups in stacks of trays, whereby a very compact structure is created. For this application the UCAO 4–9 is particularly suited (Figure 10.27).

11 Network Configurations

In this chapter typical configurations of public and private telecommunication networks including ISDN, FITL, LAN and FTTD are described.

11.1 Concept of a Telecom Administration

The telephone network of the Deutsche Telekom AG is divided into local and trunk networks (Figure 11.1). For transmission using copper technology in the local network, consisting of the intracity (metro) network and the local subscriber network, symmetrical plastic insulated twisted pair cables are in use; in the regional and national trunk networks balanced carrier frequency and coaxial cables are deployed.

The rapid development in fiber optic technology and the introduction of digital tcchnology lcd to a rcconsidcration of the traditional network concept. For this purpose, practical tests of fiber optic cable technology were carried out in all levels of the telephone network. The experimental tests took the form of competition among several companies under the auspices of the Deutsche Telekom AG.

In the following some salient data are presented, which were important for the introduction of optical fiber technology at the various network levels of the Deutsche Telekom AG.

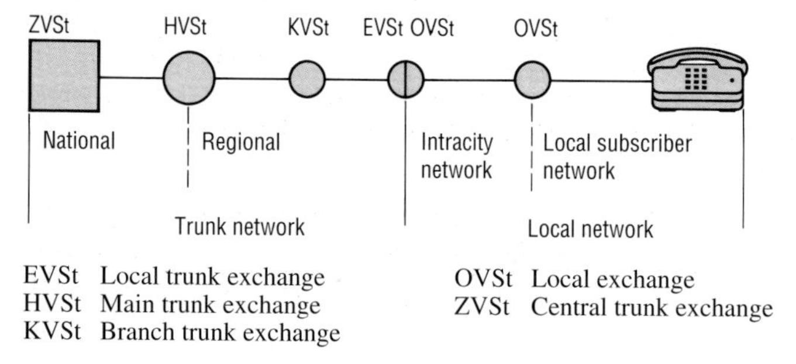

Figure 11.1
Structure of the telephone network of the Deutsche Telekom AG

Local Network

In Berlin on September 1, 1977 under the project named Berlin I, the first fiber optic local intracity trunk cable in the telephone network was hauled in. A cable 4.3 km in length with eight graded index fibers was installed; at that time the core/cladding diameter was 62.5/125 μm, with aramid yarns as strength members and a plastic sheath to protect the cable core. In March 1978 the entire plant with its 34 Mbit system at the wavelength of 850 nm was put into operation for measuring and testing purposes.

In 1980 and 1983 respectively in the projects named Berlin II and BIGFON[1)], for the first time optical fibers were run in the local subscriber network all the way into the homes of the subscribers. In both projects each subscriber terminal was connected with the exchange via two graded index optical fibers.

In 1990 the singular opportunity presented itself to test fiber optic technology in the subscriber network for the complete reconstruction of the telecommunication infrastructure in the former East German states. As part of this project named OPAL (*Op*tical *A*ccess *L*ine), in Leipzig 204 telephone lines and 37 cable television lines were installed using optical fibers in November of 1991. More than 500 000 lines in optical fiber technology in the local network of the former East German states have been employed during 1992 to 1994.

Trunk Network

In 1980 up to 480 long distance calls were transmitted simultaneously over the first optical route from Frankfurt to Oberursel which is 15.4 km in length. As a test of optical fiber technology in the national trunk network, in December 1982 the Berlin III project put a cable route 9 km in length into operation. By looping in pairs (daisy chaining) the four graded index fibers in the cable at one end point, a loop with a total length of 18 km was created for transmission of 140 Mbit/s signals. With this route length the repeater spacing distance corresponds to the existing carrier frequency routes with copper cables for the trunk network.

The next important step for the trunk network was taken in 1984 with the project Berlin IV. At that time technical solutions were presented for transmission of 140 and 565 Mbit/s signals via single-mode fibers. In this 36 km

[1)] Acronym for *b*roadband *i*ntegrated *g*lass *f*iber l*o*cal communication *n*etwork.

long cable plant, by means of wavelength division multiplexing (WDM), two 140 Mbit/s and one 565 Mbit/s signal were transmitted at the same time without regenerative repeaters over only one single-mode fiber. The WDM system was designed for wavelengths of 1300 and 1550 nm. This provided the first test for transmission at 1550 nm.

After test operation of 10 BIGFON island networks with a total of 350 subscribers in 7 cities in Germany, the Deutsche Telekom AG started to construct an interconnected optical fiber trunk network. Since 1987 all new trunk cable routes have been implemented in fiber optic technology.

In November 1987 the Deutsche Telekom AG installed the first submarine fiber optic cable in Lake Constance. Two fiber optic submarine cables with 16 single-mode fibers each connect the trunk exchanges in Friedrichshafen and Konstanz (length 21.3 km) and in Konstanz and Meersburg (length 5.9 km) by the shortest route.

Since the fall of 1989 a 9.6 km long armored fiber optic cable has been ploughed into the tidelands between the island of Föhr and the mainland of Schleswig Holstein. This initial fiber optic cable is now the middle section of a 32 km long fiber optic trunk cable between Niebüll and Wyk which has no regenerative repeaters. With this route, the versatility of modern fiber optic cables was proven and more importantly their cost effectiveness in comparison to traditional copper cable technology.

The Deutsche Telekom AG has specified its requirements for design and performance standards for fiber optic cables in Technical Terms of Delivery TS 0006 (trunk cables), TS 0009 (indoor cables) and TS 0007 (local cables). This has proven to be an important basis for general introduction of these cables. Fiber optic cable designs for the Deutsche Telekom AG are described in detail in Chapter 9.5. Depending on its structure each cable may have 2 or 4 copper pairs as order wires.

In contrast to traditional cable design, which, due to transmission technology, requires assignment of balanced low frequency, carrier frequency or coaxial pairs (structural elements) to certain levels of the network, this is not necessary for optical cable technology.

Single-fiber and multifiber loose buffers with single-mode fibers with appropriate cable designs can be used in all local and trunk networks. The single most important factor in designing such plants is the desired transmission capacity. Due to the larger repeater spacings and a higher transmission capacity of the routes, single-mode fibers are being used at all levels of the German network.

11.2 ISDN

Since the start of electrical communication transmission separate networks have been created and perfected for the special purpose of providing individual methods of communication, such as voice, text and data, corresponding to the technology possibilities. Of these, only telephone and telegraph networks have established themselves for the purpose of world-wide connections.

An international goal has now been set to build an integrated network with procedures and hardware that allows transmission not only of all traditional services but in addition of all new types of communication in one network without compatibility problems.

The application of digital technology for telephone switching and transmission is a first step. It provides the basis for an extension of the network into an "*i*ntegrated *s*ervices *d*igital *n*etwork" (ISDN). This allows transmission of voice, text, data, stationary and moving pictures (videotext, facsimile) with shorter times for making the connection, better quality and greater operational convenience.

For low frequency transmission in analog technology, only a bandwidth from 0.3 to 3.4 kHz is available. This corresponds to a bit rate of 64 kbit/s per channel in digital technology. For this transmission, the ISDN subscriber line offers a bit rate of 144 kbit/s. It is subdivided into 2 x 64 kbit/s (B channels) for voice, text and/or data and 1 x 16 kbit/s (D channel) for three possible applications (signaling, transmission of "slower" packetized data and in the future perhaps for telemetry services) (Figure 11.2).

To begin with, transmission is carried out on existing balanced copper (coaxial) cables, which makes it much easier to access the public networks.

The main difference between the traditional telephone networks and ISDN lies in the fact that, with a single telephone line access, instead of having only a telephone line with an analog voice channel the subscriber will have two digital channels available and additionally a separate 16 kbit/s channel for signaling. The present transmission network with balanced copper wires can be used for it without changes.

The international efforts for standardization were finalized during the 1984-88 ITU-T study period and most of the important ISDN standards were agreed upon. In addition ETSI has prepared European telecommunications standards, which are mandatory for European countries.

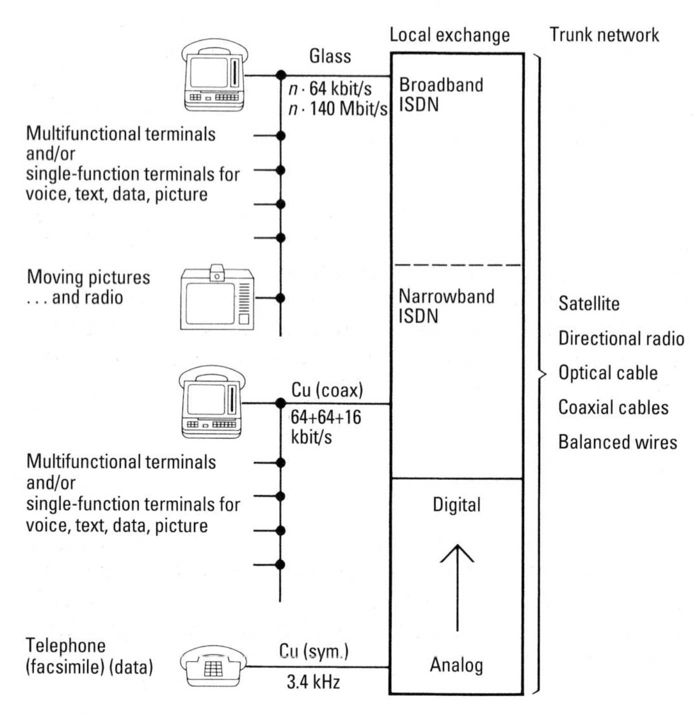

Figure 11.2
From the analog network to the broadband ISDN

11.2.1 Broadband ISDN

The concept of the Deutsche Telekom AG for further development of the telecommunication infrastructure calls for broadband services over the integrated services communication network. The standards organizations (e.g. ITU-T) are dealing with the topic of broadband ISDN. They will assure world-wide compatibility of broadband networks and services.

It is planned that the narrowband ISDN services be complemented by a variety of broadband services. These include: video telephony, video conferences, interactive videotext, fast data transmission, TV program distribution (pay-TV, high definition TV), stereo TV distribution and selection. In order to integrate these broadband services, bit rates of up to 155 Mbit/s are required for the subscriber lines – hence the requirement for optical cables.

Subscribers to broadband ISDN will only need one broadband access in order to have at their disposal, in certain gradations, the entire selection of ISDN services and levels of performance. For this purpose the basic ISDN service access (two B service channels and one D signaling channel) will be augmented by a broadband channel. Its signaling will be carried out via the D channel.

Introduction of the broadband switching technology does not require new switching systems. ISDN switching systems such as the digital electronic switch EWSD can be expanded with the corresponding modules for broadband switching. In addition to the broadband subscriber access, a broadband switching network is also being developed, which has to meet very high demands on its performance. The research and development expenditures for hardware and software are both very high.

11.3 FITL

Since fiber optic transmission technology is now in general use for trunk and intracity trunk networks of the public telecommunication operators, the local subscriber access area remains as the final network level to be upgraded. With the impetus coming from the USA, various network topologies have been suggested and studied to ascertain their technical and economical capabilities.

Depending on the end point of the fiber optic route which starts in the exchange, these *F*iber-*i*n-*t*he-*L*oop (FITL) networks, i.e. fiber optic subscriber access networks, have different names, e.g. *F*iber-*t*o-*t*he-*H*ome (FTTH), *F*iber-*t*o-*t*he-*C*urb (FTTC) or *F*iber-*t*o-*t*he-*P*edestal (FTTP). In the last two cases traditional copper cables will be connected at these end points. The final variant is particularly promising for the introduction of a fiber optic subscriber access network because costs for technical apparatus are distributed over several connected subscriber lines (laser & fiber sharing). In addition to the costs for civil works, the costs for the fiber optic plant and opto-electronic technology with lasers and photodiodes are the most important in conjunction with the local network.

Although optical fibers already offer important advantages for broadband applications today, the actual extension of fiber optic cables into the premises of individual small businesses and private customers (FTTH) will only begin when the optical solution is less expensive than the conventional copper technology, even in cases when it is used only for telephone service. In the meantime an FTTC solution oriented towards future needs is required. The *p*assive *o*ptical *n*etwork (PON) was developed for this purpose and allows use of central components such as optical fiber and electrooptic transducers by several subscribers. As the name suggests, such a network is composed of optical couplers, splitters and connectors and the fiber optic cables and splices, i.e. only passive optical components. Because of its arrangement of splitters in one or more branching elements, the structure of a PON resembles a star network (Figure 11.3).

Together the passive optical components, the central units (CU) for interactive services and for broadband distribution services, and the distant units (DU) constitute the main elements of a PON.

The CU have the task of adaptation of the PON at the interconnection point with the local exchange and the CATV headend. The central units transmit signals via a common main cable through separate optical fibers for interactive and distributive services to the passive optical couplers in the cable distribution box. These 1:n couplers split the incoming light power evenly over the n output ports. From this branching box, the signals are again transmitted via individual optical fibers in the distribution cables to the corresponding distant units. In the distant units which are located either at the curb or in the basement of a multifamily dwelling, the electrooptic conversion occurs. From there the signals are distributed to the individual subscribers via copper pairs and coaxial cables to up to 30 subscribers.

In order to implement the future FTTH solution for the PON subscribers easily and without further civil works, the subscriber drop cables are designed as hybrid cables (copper pair and optical fiber both in one cable sheath). The distant unit is the final point of the passive optical network at the subscriber end.

For the interactive services, the transmission of the signals from the CU to the individual DU, i.e. towards the subscribers, is carried out in the time division multiplex technique (TDM). In the opposite direction transmission of information is more difficult, because the differing geographic location of the subscribers results in differing delay times. All subscribers must on the other hand be able to access the main cable between the passive optical coupler in the cable branching box and the central unit via the same optical fiber. For transmission in this direction the multiplex technique TDMA with multiple access, also in use in satellite technology, is required. It controls the timing of the access of each subscriber in order to assure that the signals arrive at the CU in order with respect to delay times.

The medium of the PON is the single-mode optical fiber. Due to cost considerations interactive services are carried out in both directions on each optical fiber, i.e. bidirectionally, using 1300 nm in the direction towards the exchange and in the opposite direction towards the subscriber at 1550 nm. The optical filter which is required to separate the directions has only low

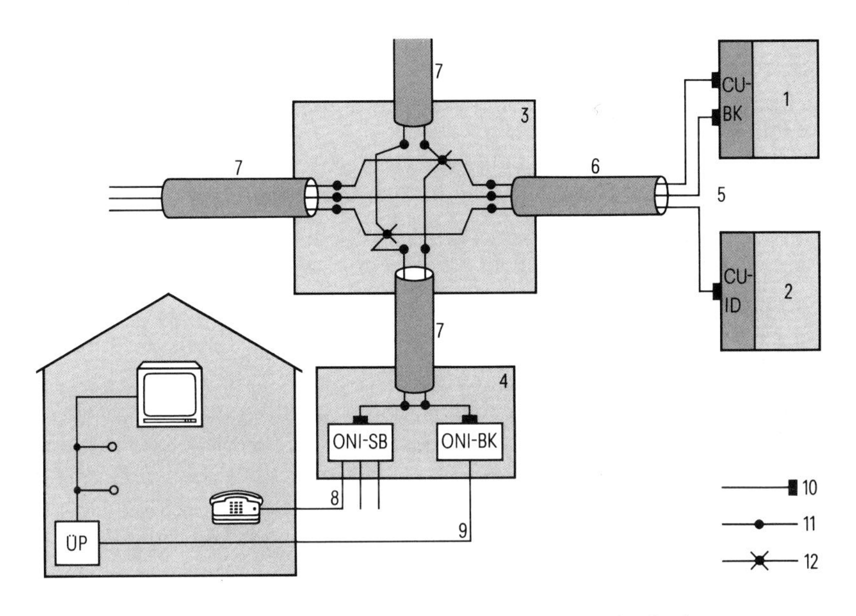

CU-BK	Central unit for broadband communication
CU-ID	Central unit for interactive services
ONI-SB	Optical network interface for narrow-band services
ONI-BK	Optical network interface for broadband communication
ÜP	Interconnection point

1 BK headend
2 Local exchange
3 Cable distribution box
4 Distant unit (DU)
5 Central unit (CU)
6 Fiber optic main cable
7 Fiber optic distribution cable
8 Symmetric copper cable
9 Coaxial copper cable
10 Fiber optic connector
11 Fiber optic splice
12 Fiber optic coupler

Figure 11.3
Passive optical network (PON)

transmission loss. Therefore the distance penalty is relatively low in comparison to a solution with two separate optical fibers. The range depends mainly on the transmission bandwidth chosen, i.e. on the number of subscribers and on the type of service to be transmitted. However, the number of splices and connectors and the attenuation coefficient of the optical fiber also influence the attenuation budget. Due to the delay difference to be compensated for by the TDMA method the range for interactive services is limited to 10 km. The range for optical TV transmission is about 9 km. Because the majority (90%) of all subscriber lines are shorter than 4.5 km, there are sufficient system margins available for transmission of both services.

Both the active network termination devices in the form of the DU and the subscriber telephones which previously were served via copper pairs from the exchange battery need a fail-safe power source. Since no preexisting solution is available which assures sufficient energy in light form for transmission via fiber to the subscriber and storage there, the problem of power supply in optical local access networks must be reconsidered.

Three alternatives for power supply of the subscriber lines are possible: from the exchange, locally and from the subscriber side.

When the distance between the exchange and the DU is short, a remote powering from the exchange power supply is possible when the diameter of the conductor which runs parallel to the fiber optic cable is not too large.

In the case of local power supply to the DU with a direct connection to the public power network, an additional back-up battery is needed for use during unavoidable power interruptions. This is also needed in cases where operation is served via the private power network of the subscriber. These batteries must assure operation of the interactive services for at least several hours.

Summing up, it can be said, that none of these three possibilities is an ideal solution for the power supply; a selection will have to be made based specifically on the local network situation.

11.4 LAN

The term *l*ocal *a*rea *n*etworks (LAN) was introduced to differentiate between processor links with very high bit rates over very short distances on the one hand and long distance networks with relatively low bit rates but very long route lengths on the other hand. The individual stations of a LAN are usually between 100 m and several kilometers from one another whereby the bit rates can reach up to several 100 Mbit/s.

The following definition for local area networks was worked out by the working group TC 97 of the International Organization for Standardization:

"A local area network is a network for bit-serial transmission of information between interconnected independent terminals. The user has full responsibility for it and it is limited to his property."

In Germany special permission from the Deutsche Telekom AG is needed for interconnection of buildings on different plots of land.

In general LAN systems are differentiated according to their topology, their applications criteria, their interfaces and access methods, as well as according to their transmission media. In contrast to most applications in analog and digital technology, information is usually distributed in data packets and not via firmly assigned channels such as a telephone connection, for which a transmission channel is permanently switched for the duration of the call.

LAN applications are mainly in the area of office communications (electronic mail), personal computer (PC) networks and in industry (e.g. for process control).

For LAN outside of the campus area of telephone networks, separate fiber optic cable networks are necessary.

Interconnection, e.g. between the ISDN and LAN communication networks, can be accomplished through gateways, which provide access to the public networks for the LAN.

The choice of transmission media (copper wires and/or optical fibers) depends mainly on the technical requirements and economic efficiency of the system in question. Optical fibers are particularly advantageous for LAN applications due to their low attenuation, high bandwidth, galvanic separation of the subscribers, high safety from interception and insensitivity to electrical disturbances.

Depending on the importance of these characteristics, there are certain applications with high bit rates that can only be met by optical fiber technology, e.g. bank and insurance networks, university networks, networks for medical applications, in CAD processes and networks in endangered area, in military networks and networks for niche markets.

11.4.1 Networks

In LAN topologies (the totality of the transmission paths) three network types are distinguished: star, ring and bus structures.

Star Networks

For the reasons listed above fiber optic cables are particularly useful for star network configurations (Figure 11.4). The basic specification is laid out according to Ethernet[1]).

Typical values are 10 Mbit/s with the access method CSMA/CD (*c*arrier *s*ense *m*ultiple *a*ccess with *c*ollision *d*etection, which is also described in IEEE 802.3 standard).

[1]) Local data network, specified in the IEEE 802.3 standard.

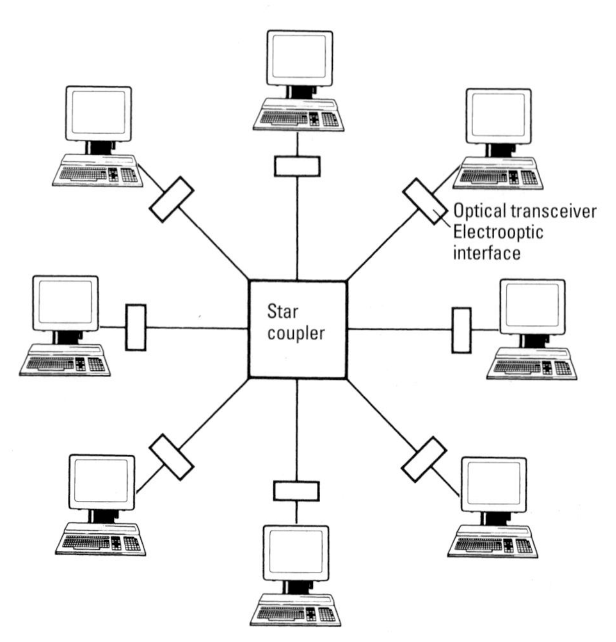

Figure 11.4 Schematic of a star system

In the center points of such networks there are active star couplers; from them optical cables lead either to the corresponding terminals or to the next star coupler.

Star networks offer a topology which is independent of the services carried and is flexible, permitting port-related implementation of a large number of data link functions.

Ring Networks

Optical cables can be used widely in ring networks. The signals are transmitted from station to station, either until they reach the receiver station or until the data package arrives back at the point of origination. Each time the signal is transmitted, it is amplified (Figure 11.5). Token access has established itself as a method of allotment of the right of transmission (standardized in IEEE 802.5). A token is passed around the ring; possession of the token gives the right to transmit a communication package.

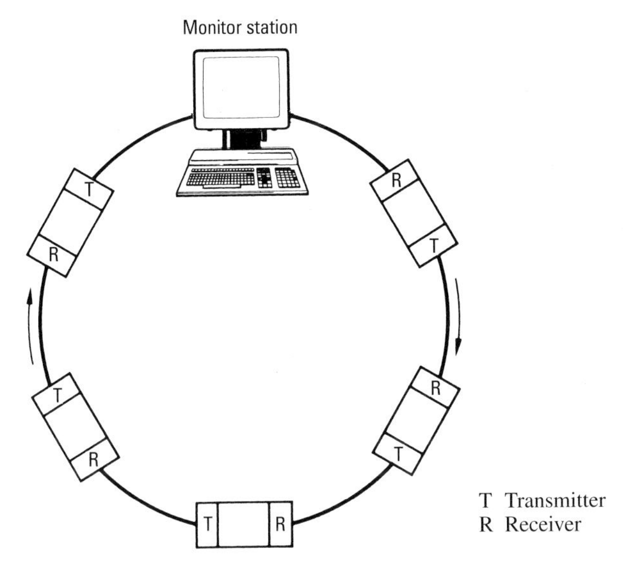

Figure 11.5 Schematic of a ring system

Bus Networks

Baseband systems, specified according to Ethernet, have already established themselves for a great many applications. In bus systems (Figure 11.6), special balanced as well as coaxial copper cables and, to lengthen routes, fiber optic cables are used as transmission media. The data are transmitted via a passive hook-up on the coaxial cable in both directions and deleted by non-reflecting terminal resistors at both ends. These data packages carry receiver and transmitter addresses and are read by the addressed station(s). The access method in office systems is the CSMA/CD system discussed below. In industrial applications, the token bus is used (standardized in IEEE 802.4).

Before a station begins transmission, it "listens" to find out if signals are already being transmitted. If this is the case, it waits until the end of that transmission; otherwise it begins transmitting. As transmission begins, the station "listens" to the transmission channel. If it perceives a collision with data of another station, which began transmitting at about the same time, then it interrupts transmission and repeats it at a later time, determined by a random number generator. This system assures that all stations will have equal access to the transmission medium, independent of their geographic configuration.

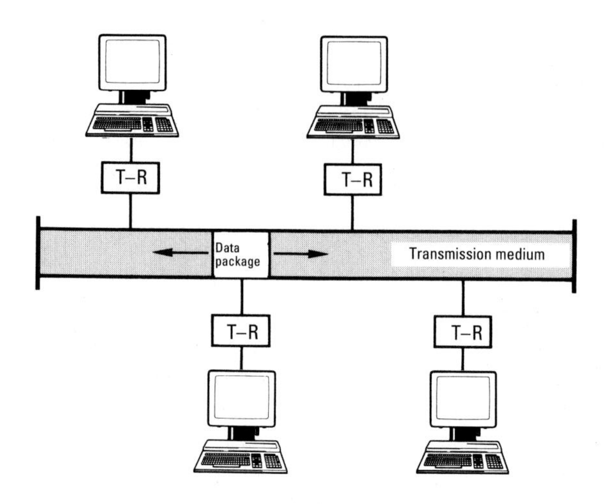

Another frequent application of such networks lies in linking personal computers, which via a LAN gives them access to common memories, printers and processing units.

11.4.2 Cabling Concepts

In the past, cabling inside buildings and in the outdoors was planned according to the services to be provided (voice, data). The result was that often a number of different fiber optic and copper cables were installed parallel to one another to meet special implementation needs, which did not meet later the demands of the modern communications world. Not enough attention is given to consideration of the high mid-term additional costs of extensive storage of materials, more difficult documentation of the plant and servicing as well as initial training and training upgrades for personnel when calculating the imagined lower costs of inexpensive initial investment.

Structured cabling concepts in contrast offer a solution for a future oriented application independent of the service offered. They conform to standards, are flexible and sufficient to meet highest standards of communications technology in all elements, e.g. cable, terminal technology, ease of installation.

Siemens developed the structured cabling system ICCS (*i*ntegrated *c*ommunications *c*abling *s*ystem). In it a distinction is made between the primary (backbone, campus), the secondary (backbone, building and rising areas) and the tertiary area (subscriber, individual floor) (Figure 11.7).

In primary and secondary areas, mainly fiber optic cables are used. In the tertiary area mainly balanced shielded copper cables or alternatively fiber optic cables are implemented; this means that only two types of cables are required.

11.4.3 FDDI

Development of new technologies in electronics products for the communications sector and continuous improvement of fiber optic technology has made it possible to design local networks for high value applications.

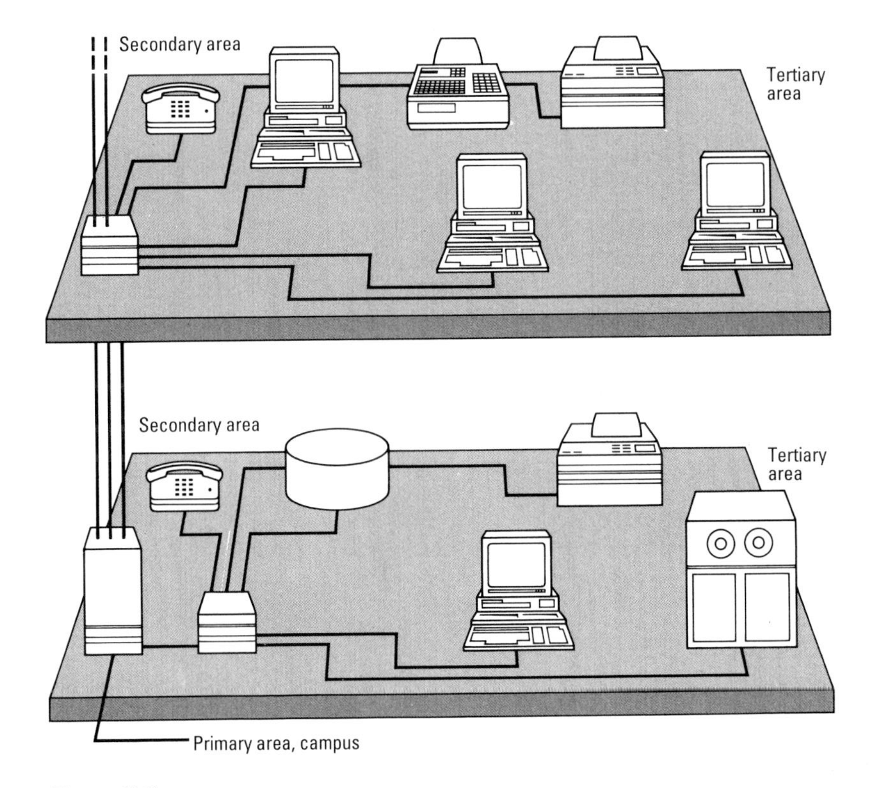

Figure 11.7
Structured cabling system ICCS with only 2 cable types
Primary area: fiber optic cables, Secondary area: fiber optic cables and ICCS cables.
Tertiary area: ICCS cables.

FDDI (*f*iber *d*istributed *d*ata *i*nterface) is a token passing technique conceived for the fiber optic transmission medium. In contrast to Ethernet (10 Mbit/s) and Token Ring (4 and 16 Mbit/s) it permits transmission speeds of 100 Mbit/s and is therefore a high speed fiber optic network.

The following characteristics are typical of this high speed technique:

▷ Use of fiber optic technology throughout, combining exceptional transmission characteristics with minimum interference susceptibility,

▷ nominal data rate of 100 Mbit/s with sufficient capacity for backbone and front end applications,

▷ option of up to 1000 physical connections over a fiber length of 200 km,

▷ excellent throughput characteristics in the high load and overload range and impartial control of network access through the use of the Token access principle,

▷ fault tolerant design including two counterrotating fiber optic rings and an intelligent station management as well as

▷ optional use of copper cabling in the subscriber access area.

Therefore FDDI can interconnect island solutions, which were frequently installed in the past and are based on Ethernet (10 Mbit/s) or Token Ring (4/16 Mbit/s), to create a complete network where desired, as well as so-called front end solutions (100 Mbit/s to the work station).

In the campus area (i.e. between buildings, the primary area) an FDDI dual ring is constructed as a network over all the buildings. The stations in the dual ring (mainframes, bridges/routers or concentrators), the so-called dual attachment stations (DAS) or Class A stations, are connected with both fiber optic rings and therefore remain operational even when one connection fails.

The connection to the floor distributor stations (secondary level) and the terminal equipment (tertiary level) is achieved by a tree like structure. The connection to these "trees", so-called single attachment stations (SAS) or Class B stations have direct access to only the primary ring (Figure 11.8).

In this way it is possible to increase the fault tolerance of the overall network since, in the event of a Class B station failing, the dual ring remains fully operational. The small number of Class A stations and optional protection by auxiliary equipment, such as redundant power supply or an optical bypass, reduce the probability of the FDDI ring splitting into two separate rings if two DASs fail.

The transition to the subnetworks in the individual buildings or on individual floors is achieved by bridges or routers. They make it possible to intercon-

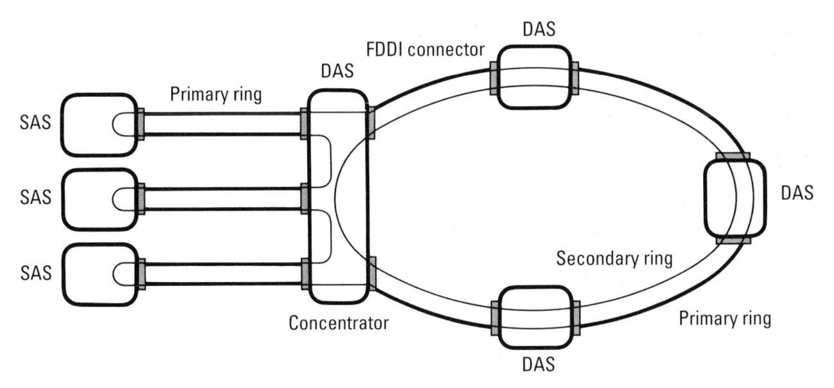

Figure 11.8 FDDI topology

nect local Ethernet or Token Ring networks and are also available as interfaces for WAN (wide area networks) connections.

In addition to applications as a backbone network, FDDI will also become established as a front end solution in the future as new generations of computers and applications of the nineties require data rates which cannot be handled by any of the conventional LAN technologies. For these applications today, it is already possible to use high efficiency copper cabling (e.g. ICCS = Integrated Communication Cabling System) in addition to fiber optic technology in the tertiary level.

Modular concentrators are becoming increasingly important as active FDDI components. They not only allow connections to an FDDI ring (Class A and Class B), but also offer high flexibility in the configuration and the internetworking of local area networks by integrating different LAN technologies. An integrated network management permits the supervision and configuration of the entire network from one station.

11.4.4 MAN and Outlook

In the past, local area networks were connected over a wide range via 9.6, 64 or 2048 kbit/s dedicated circuits of X.25 packet connections. The need arose for a fast and cost effective overall communications network which could handle both the data transport and the transmission of voice applications. In recent years the MAN (*m*etropolitan *a*rea *n*etworks) have been conceived to meet these needs for network access, transmission and exchange, which employs the DQDB (*d*istributed *q*ueue *d*ual *b*us) technique.

MAN based on DQDB can be described as having the following characteristics (Figure 11.9):

▷ Wide area (> 100 km),

▷ high bit rates (34/45/140 Mbit/s),

▷ basic structure: 2 independent, counter-rotating bus systems,

▷ cell relay transmission structure,

▷ up to 255 nodes and a maximum of 4000 subscribers,

▷ dynamic bandwidth assignment,

▷ simultaneous transmission of voice and data

▷ service integration in conjunction with broadband ISDN.

The MAN network topology for a public network is composed of two similar network levels, which are ordered in hierarchy. The upper level of the hierarchy is created through a bus, a daisy chained bus or a point to point connection.

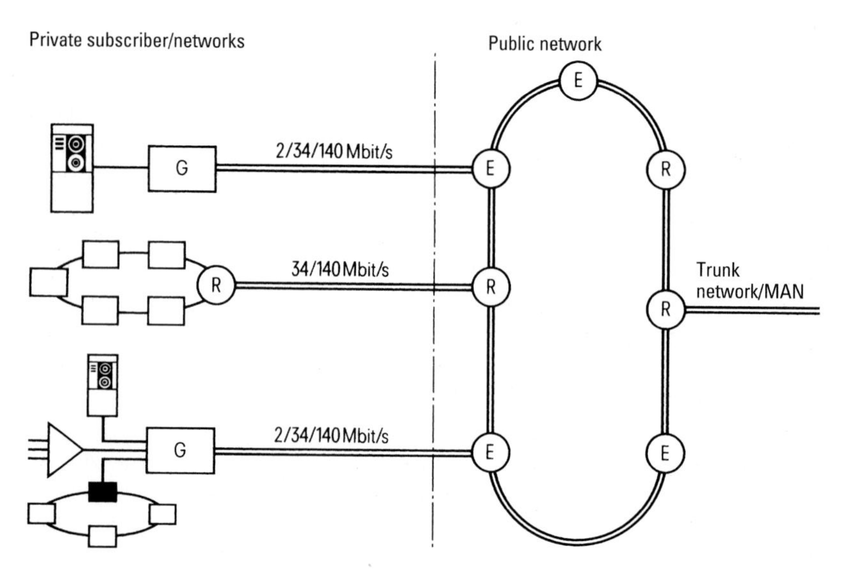

Figure 11.9 MAN-Structure (G gateway, E edge, R router)

To this network, which is the actual component of the MAN, the so-called edges are connected. They are the access point of the lower level of the hierarchy and in it they provide access to the subscriber networks CAN (*c*ustomer *a*ccess *n*etworks) by means of structures like those of the upper level. The nodes of the CAN, so-called customer gateways, are located on the premises of the customer. LAN, private branch exchanges and multiplexers can be connected to them.

The control and supervision of a MAN is carried out as for FDDI from a central management station.

The MANs gain their important place in today's communications world due to their integration of the future oriented broadband ISDN. B-ISDN is based on the asynchronous transfer mode (ATM). In the metropolitan area networks transmission is carried out without fixed connections (datagram structure), whereas ATM transmission is oriented towards connections and carried out via logical channels, i.e. before the transmission of information a definite connection is set up. Because the cell structure, in which the data are packetized for transmission, is the same for MAN and ATM techniques, both services can be seamlessly connected by a so-called connectionless server. This means that the cost efficient, physical migration from MAN to future network structures based on DQDB is possible. Furthermore, the fast pace of development in the microelectronic sector will permit future ATM systems with frequencies in the Gbit/s range and allow the present communications world to move into unimagined dimensions.

11.5 FTTD

Fiber-to-the-desk (FTTD) is quickly becoming an accepted part of the communication infrastructure in most modern offices, especially now that standards have been set for fiber optic applications and passive network components. For instance, the structured cabling standard ISO/IEC 11801 defines applications for Ethernet, Token Ring, and fiber distributed data interface (FDDI) based on optical fibers. In addition, the ATM standard (155 Mbit/s) is in the process of being adopted, while a 622 Mbit/s standard is under discussion. Once these standards have been established, demand for fiber optic technology is certain to surge.

At the same time, the length of the floor cabling is no longer limited to the once strict requirement of 90 meters, and a new 300 meter building zone is being defined. Within this area all applications can be transmitted without repeaters over a single medium: Fiber optic cable with multimode fibers (Table 11.1). This facilitates the migration to emerging ATM and multimedia technologies.

Application	Bandwidth (Mbit/s)	Floor < 100 m		Building < 300 m		Campus < 2000 m	
		Medium	Technology	Medium	Technology	Medium	Technology
10 Base-F	10	MM	S	MM	S	MM	S
Token Ring	16	MM	S	MM	S	MM	S
100 Base-F	100	MM	S	MM	S	MM	LE
FDDI	100	MM	S	MM	S	MM	LE
ATM/STM	52	MM	S	MM	S	MM	LE
	155	MM	S/LE	MM	S/LE	MM	SL/LE
	622	MM	SL/LE	MM	SL/LE	SM	LL
	1244	MM	SL	MM	SL	SM	LL
	2488	MM	SL	MM	SL	SM	LL

ATM Asynchronous transfer mode
FDDI Fiber-distributed data interface
LE Light emitting diode (1300 nm)
LL Laser (1300 nm)
MM Multimode (62.5/125 µm)
S LED (850 nm)
SL Laser (850 nm)
SM Single-mode (9/125 µm)
STM Synchronous transfer mode

Table 11.1
Recommendations for building cabling: Cabling zones and technology

Greater network flexibility and transparency

On the whole, conventionally structured cabling is described in terms of riser and horizontal subsystems. It has been widely held that fiber optic cabling is not needed in the last 90 m of the horizontal section. In particular now that ISO has laid down standards for data transmission capacities via copper cables, for example category 5, referring to a capacity of 100 Mhz. However, this would unnecessarily restrict the type of network design users could plan. Fiber optic cabling can overcome this division between horizontal and vertical cabling, since multimode fibers – with bandwidth to support up to 2.5 Gbit/s and with loss of less than 0.3 dB over a distance of 300 meters – are ideally suited for cabling within buildings. Thus networks can be structured simply, flexibly and transparently. Three concepts have been met with particularly wide acceptance: Centralized cabling, open office cabling and zone cabling.

Centralized cabling

With centralized cabling, all network users within a building are administered from a central data room (Figure 11.10a). Workstations can be up to 300 m away from the hub, so in more than 95% of all buildings one central

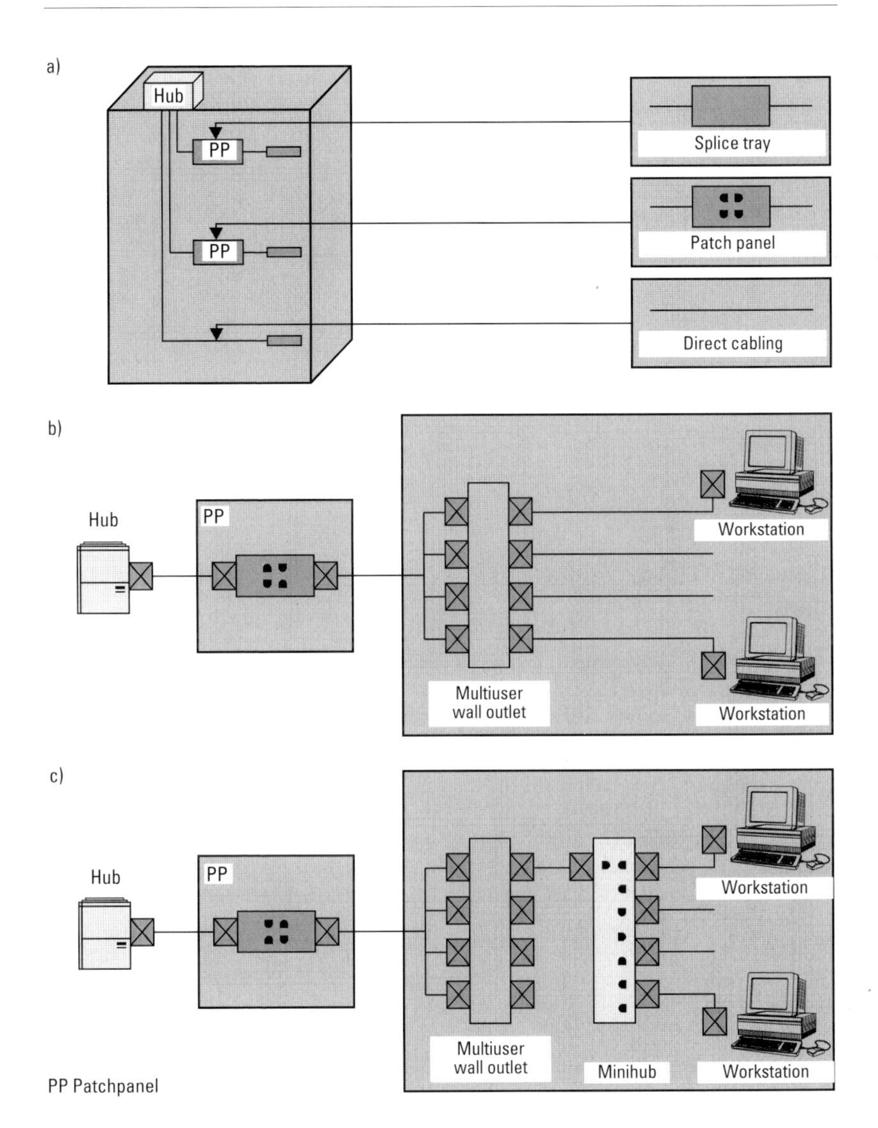

Figure 11.10 Cabling concepts for network structures

a) Centralized cabling
b) Open office cabling
c) Zone cabling

data room can adequately administer the entire local network. Cables can either be spliced, interconnected in the telecommunication distribution panel or installed directly from the hub. Instead of additional electronics, telecommunication distribution panels and communication rooms on individual floors, a passive patch panel is all that is needed. Smaller and less complex, centralized cabling takes up less office space and may result in lower operating costs for air-conditioning and electricity, for example.

This architecture can even lead to lower cabling and electronics costs since the ports of active components are used more efficiently. In addition, centralized cabling offers a higher level of network flexibility and is less expensive to administer and maintain. Instead of having to move electronics and workstations, workstations can be assigned to specific workgroups by simply patching connecting cables at the main distribution frame. Thanks to the relatively straight-forward network infrastructure, errors can be more easily pinpointed and eliminated.

Open Office Cabling

A centralized cabling architecture combined with multiuser wall outlets – normally connecting up to 12 users (Figure 11.10b) – is particularly well-suited for open offices. Since only one cable is horizontally routed to the multiuser wall outlet, installation costs are lower and the cable duct ultimately takes up much less space. Connecting cables are routed from the multiuser outlet, through the partition walls, to the individual workstations. Since fiber optic connecting cables are not susceptible to electromagnetic compatibility problems, power supply cables routed through partition walls to the workstations do not hamper their performance. The length of fiber optic connecting cables also has no bearing on transmission quality for these distances. If the partition walls in an office end up being removed, the connecting cables may need to be reconfigured, but no new horizontal cabling will be required.

Zone Cabling

In this architecture, the network cards that exist in the workstations for the copper network are integrated into the fiber optic network. The entire passive network is built with centralized cabling and fiber optic multiuser outlets (Figure 11.10c). A minihub is installed at one port to serve as the bridge between the optical fiber and the copper network card. The workstations can then be connected to the fiber optic network via this minihub. If the copper network card is eventually replaced by fiber network cards (for example, ATM network cards), only the minihub is removed and the workstations can be directly connected to the multiuser outlet. Since the fiber optic network has already been installed, recabling is not necessary. In this respect, zone cabling safeguards the migration to high-speed networks without jeopardizing previous investments in existing copper network cards.

12 Electrooptic Signal Conversion

In order to transmit light signals in an optical fiber, suitable transmitter and receiver elements are needed at the beginning and end of the optical fiber to convert electrical signals into optical ones, and vice versa (Figure 12.1). At the transmitter side, an electrical signal modulates the intensity of a light source. The optical signal is coupled into the optical fiber and arrives at the receiver; there a photoreceiver reconverts it into an electrical signal.

For electrooptic transducers used in semiconductor technology chemical elements of the third (Al, Ga, In), fourth (Si, Ge) and fifth (P, As) group of the periodic system have proven to be very useful (Table 3.1). The alloys of these III, IV and V semiconductors are particularly important and of these mainly the quaternary (fourfold) compounds InGaAsP and GaAlAsP and their ternary (threefold) compounds InGaAs and GaAlAs which can be applied to the substrate InP or GaAs by matching lattices, i.e. due to the fact that both crystals have the same lattice constants.

Semiconductors have two energy ranges for electrons, namely the *valence band* and the *conduction band*, which are separated from each other by the energy gap E_g (Figure 12.2).

Furthermore, the composition of these compounds is chosen to achieve the optimum band gaps and refractive indices needed for the functions of the components. The exact thickness of the layers and proper homogeneity is produced through the epitaxy method (regularly aligned crystal growth). Three methods are differentiated: Liquid phase epitaxy (LPE), metal oxide vapor phase epitaxy (MOVPE) and the molecular beam epitaxy (MBE/MOMBE) of which the latter two methods are usually applied.

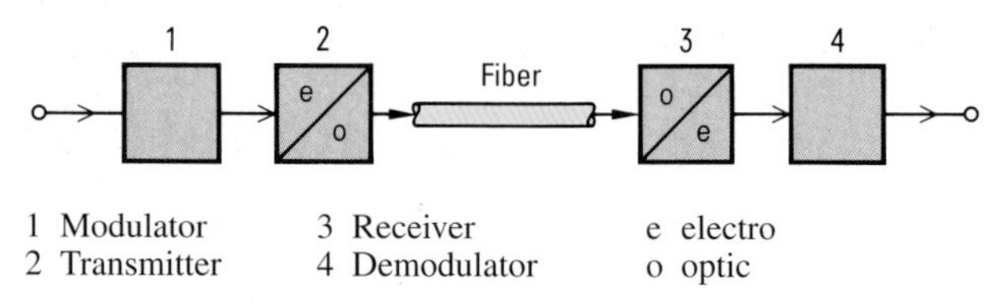

1 Modulator
2 Transmitter
3 Receiver
4 Demodulator
e electro
o optic

Figure 12.1 Schematic of an optical fiber transmission system

A photon radiated into the semiconductor releases its energy $h \cdot \nu$ ($h = 6.62608 \cdot 10^{-34}$ Js is Planck's constant and ν is the frequency of the light) to an electron in the valence band, thereby lifting it into the higher energy conduction band. The photon disappears in the process and the electron leaves an unoccupied spot in the valence band, known as a positive hole. This is one of the electrooptic interactions – the absorption.

If the conduction band has more than an equilibrium distribution of electrons, then electrons can fall back spontaneously into holes of the valence band. Thereby one photon is radiated out of the semiconductor for each electron. This process is known as radiative recombination of excess charge carriers, because in the process, extra electrons and holes unite. It is called spontaneous emission or luminescence.

Stimulated emissions occur when photons in the semiconductor cause excess charge carriers to recombine radiantly, i.e. excite the emission of photons. Of importance in this case is the fact that the radiation emitted is identical in phase and wavelength to the radiation exciting it.

These three processes always occur simultaneously when one of them predominates and can be put to use technically. The photodiode makes use of absorption, the light emitting diode uses spontaneous emission and the laser diode utilizes stimulated emission.

It is also important to note that semiconductors, in which the electric current is conducted by electrons (*n*egatively charged particles) are called *n-type semiconductors*. In contrast, a semiconductor with few electrons, in which mainly the holes (*p*ositively charged carriers) conduct electricity, are called *p-type semiconductors*. A combination of two layers with a p-type semiconductor and an n-type semiconductor, i.e. a *pn junction*, makes up a diode.

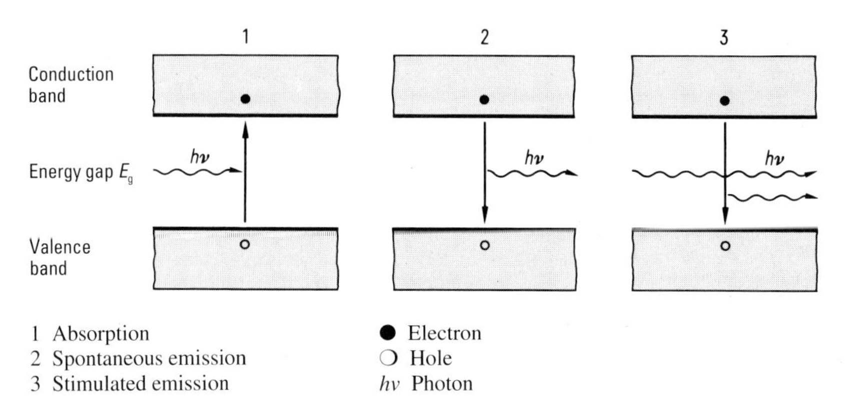

Figure 12.2 Electrooptic interactions in a semiconductor

12.1 Transmitters

In order to obtain spontaneous or stimulated emission to radiate photons, excess charge carriers must be introduced into the semiconductor. This is accomplished by injecting charge carriers via a pn junction. If the pn junction is operated in the forward direction, additional electrons are injected into the p-layer and additional holes into the n-layer, which can be "used" to radiate photons.

The process of injection of charge carriers together with the emission that follows is called *injection luminescence*. It finds application in transmitters (emitters), e.g. LEDs and laser diodes.

12.1.1 LED

A semiconductor diode which radiates light by spontaneous emission is called a light emitting diode (LED).

The quality of the conversion of electrical current into light is described by the external *quantum efficiency*, which describes the number of photons emitted per unit of time relative to the number of charge carriers crossing the pn junction in the semiconductor diode (for GaAs it is typically 0.5 to 1.0 per cent).

Because the quantum efficiency decreases as the temperature increases, a heating of the recombination zone must be avoided, i.e. a corresponding dissipation of heat must be assured, e.g. through a suitable construction of the LED.

The wavelength of the light emitted is also important for the operation of the LED. It is determined mainly by the energy gap E_g, whereby the following relation applies:

$$\lambda = \frac{h \cdot c}{E_g}$$

λ wavelength in µm
E_g energy gap in eV
$h \cdot c = 1.2398$
$1 \text{eV} = 1.60218 \cdot 10^{-19} \text{J}$

For the value of E_g in gallium arsenide (GaAs) diodes, the value 1.43 eV applies and therefore for λ the value is 0.87 µm (at room temperature). For indium phosphide (InP) with E_g = 1.35 eV, the corresponding value is $\lambda = 0.92$ µm.

The spectral width $\Delta\lambda$ of an LED is approximately proportional to the square of the wavelength λ; therefore it increases greatly toward longer wavelengths.

The response time of an LED is a further important characteristic. It is related to the rise time of the diode. The average lifetime of the excess charge carriers is namely the determining factor in light emission after the current is turned off. Typical minimum lifetimes lie in the range of a few nanoseconds, which corresponds to a modulation bandwidth of the order of 100 MHz.

Because rise time and quantum efficiency depend on the concentration of holes, they cannot both be optimized at one time. Particularly "fast" LEDs emit, relative to the injection current, fewer photons.

Structure and Characteristics

In the following, three examples of LEDs are considered as examples for the many possibilities. They are:

Planar GaAs diode for $\lambda \approx 900$ nm.
High radiance AlGaAs/GaAs diode for $\lambda \approx 830$ nm.
High radiance InGaAsP/InP diodes for $\lambda \approx 1310$ nm.

In Table 12.1 typical values are listed.

GaAs Diode for Wavelengths around 900 nm

The GaAs diode has the simplest structure of the transmitter diodes (Figure 12.3). The diode is composed of n-GaAs substrate with a p-area 200 μm thick on the top side. The part of the infrared light generated mainly in the p-area, which is emitted on the p-surface, is used for launching into the optical fiber.

High Radiance AlGaAs/GaAs Diode (Burrus Type[1)]) for Wavelength around 830 nm

The diode is composed of a cuboid semiconductor crystal. A double hetero-structure composed of three AlGaAs layers with different thickness and dopings is grown epitaxially[2)] on an n-doped GaAs substrate. With these epitaxial layers facing downward and an intervening gold stud as a heat sink, the diode is soldered to a silicon chip with a conductor contact and SiO_2

[1)] First proposal by C.A. Burrus for this diode structure.
[2)] Epitaxy: regularly aligned crystal growth

Table 12.1 Characteristics of LEDs

		Type		
		LED	High radiance LED	High radiance LED
Wavelength	nm	900	830	1310
Spectral Width	nm	40	40	120
Semiconductor material		GaAs	AlGaAs/GaAs	InGaAsP/InP
Structure		Planar	Double heterostructure	Double heterostructure
Emission		Spontaneous	Spontaneous	Spontaneous
Light power launched into a fiber with core diameter of 50 μm	μW	5	20	20
Maximum bit rate in	Mbit/s	5	60	100

insulation. On the underside of the diode crystal, an Al_2O_3 insulation layer limits the flow of current to the p-contact with its small surface area.

The thickness and in particular the doping level of the AlGaAs covering layer above it are chosen so that the current does not measurably dissipate. The result is that the active AlGaAs covering layer is excited to emission within an area of only slightly larger diameter than that of the p-contact, thus assuring the desired small emitting area (Figure 12.4).

The infrared light radiated upward toward the GaAs substrate is utilized for launching into the optical fiber. This radiation is only negligibly self-absorbed in the active layer, because this layer is very thin, and passes

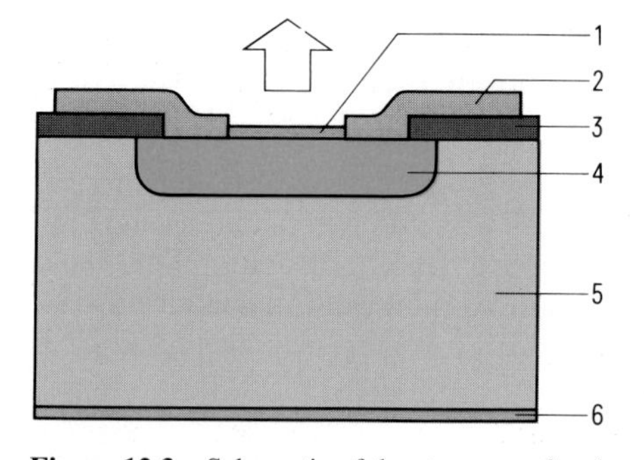

1 Antireflection coating
2 p-contact
3 Al_2O_3 insulation
4 p-GaAs (diffused)
5 n-GaAs (substrate)
6 n-contact

Figure 12.3 Schematic of the structure of a planar GaAs diode

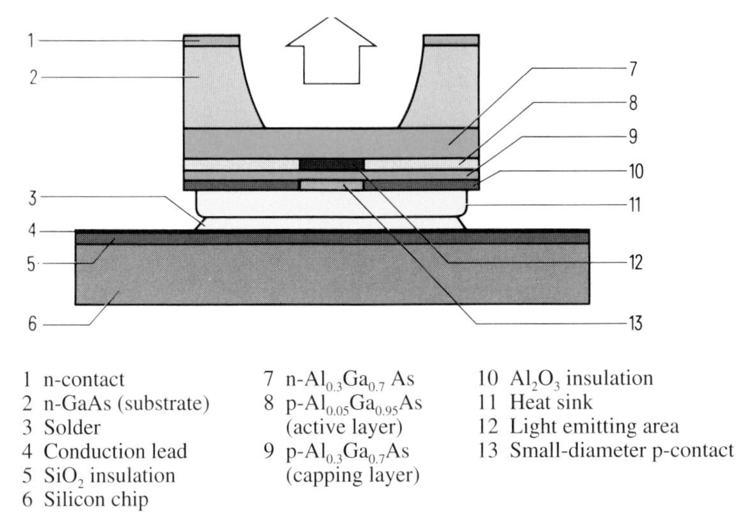

Figure 12.4 Schematic of the structure of the high radiance AlGaAs/GaAs diode

through the AlGaAs layer above it without absorption, although it would be completely absorbed in the GaAs substrate within only a few micrometers. Therefore, a well which is aligned concentrically over the p-contact is etched in the GaAs substrate above the emission region. The n-contact is applied to the remaining GaAs material. This construction has the advantage that, with the aid of the p-contact with its small surface area, a correspondingly small light emitting surface can be created.

High Radiance InGaAsP/InP Diodes for Wavelengths around 1310 nm

For transmitter diodes for the "second window" (1300 nm), a change in materials must be made from the ternary semiconductor AlGaAs to the quaternary semiconductor InGaAsP and to the InP substrate crystals. Basically, the structure of quarternary high radiance diodes (Figure 12.5) corresponds to that of the AlGaAs/GaAs diodes.

A semiconductor made of InGaAsP emits at a wavelength of 1300 nm and is lattice matched to InP. Therefore the double heterostructure which is highly effective for transmitter diodes can again be realized, in which the InP takes the role of the "barrier". A p-doped layer of InGaAsP functions as the contact layer. The substrate material takes the form of an integrated lens to improve the launching of light into the optical fibers.

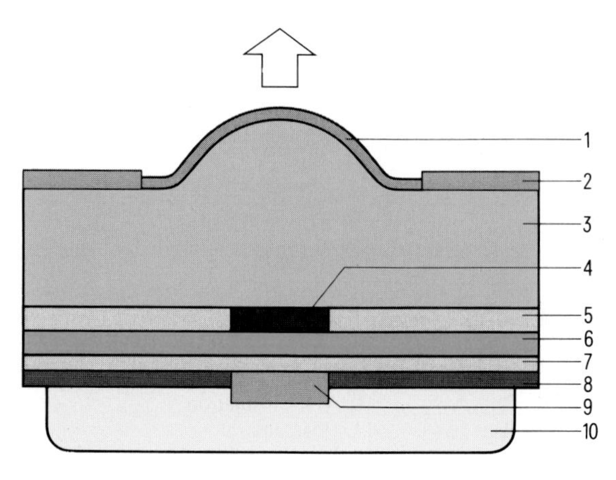

1 Antireflection coating
2 n-contact
3 n-InP (substrate)
4 Light emitting area (diameter ∅ 25 to 30 μm)
5 InGaAsP
6 p-InP
7 p-InGaAsP
8 Al_2O_3-insulation
9 p-contact (diameter 25 to 30 μm)
10 Heat sink

Figure 12.5 Schematic of the structure of the high radiance InGaAsP/InP diode

12.1.2 Laser Diode

A laser diode is a source of radiation, which utilizes stimulated emission (see Figure 12.2). Laser is the abbreviation for *l*ight *a*mplification *b*y *s*timulated *e*mission of *r*adiation. Through high current density, a large excess of charge carriers is generated in the conduction band in the laser so that a strong stimulated emission can take place. This amplification effect due to an avalanche of photons is enhanced with an optic resonator, which is usually composed of two plane parallel partially transparent mirrors. These mirror surfaces in the laser diode are natural crystal surfaces created by cleavage of the semiconductor crystal and provided with an additional protective layer.

In order to show the difference between an LED and a laser diode, Figure 12.6 shows a typical light current characteristic curve. As diode current is increased, a threshold is reached at which light amplification in the crystal averages out losses due to attenuation and radiation. Above this threshold, strong laser emission begins. In contrast to LEDs with their broad spectral distribution, the emission in laser operation narrows to one or only a few spectral lines (Figure 12.7).

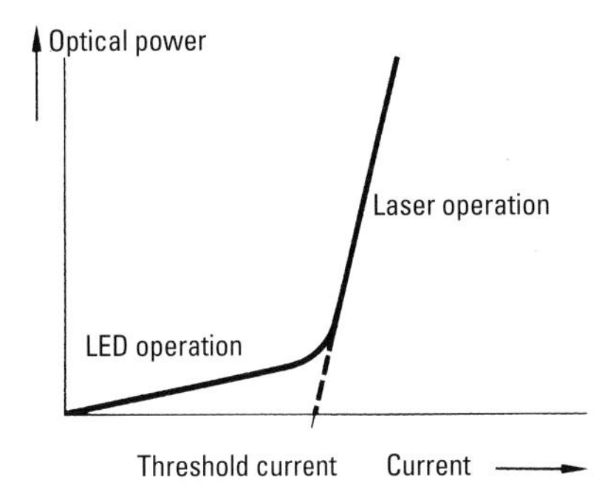

Figure 12.6 Light-current characteristic of a laser diode

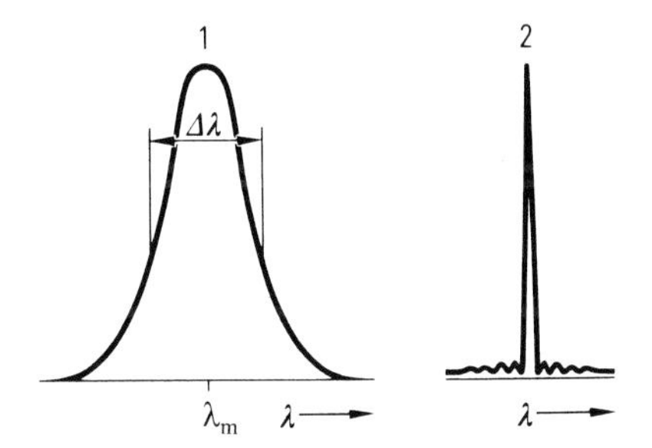

Figure 12.7 Spectral distribution of the emission of (1) an LED and (2) a laser diode

The width of a transmitter spectrum is usually given as the full width at half maximum (FWHM) $\Delta\lambda$ or as the full root mean square width (FRMS) $\Delta\lambda_{\mathrm{FRMS}}$, whereby, for a Gaussian-shaped spectrum, the following relation applies:

$$\Delta\lambda = \sqrt{\ln 4}\ \Delta\lambda_{\mathrm{FRMS}} \approx 1.18 \cdot \Delta\lambda_{\mathrm{FRMS}} \text{ or}$$

$$\Delta\lambda_{\mathrm{FRMS}} \approx 0.85 \cdot \Delta\lambda.$$

In contrast to the emission of an LED, the radiation of a laser diode is spatially coherent due to stimulated emission. The radiation beam is noticeably narrower than that of the LED, which facilitates a particularly effective light launching into the optical fiber (Figure 12.8).

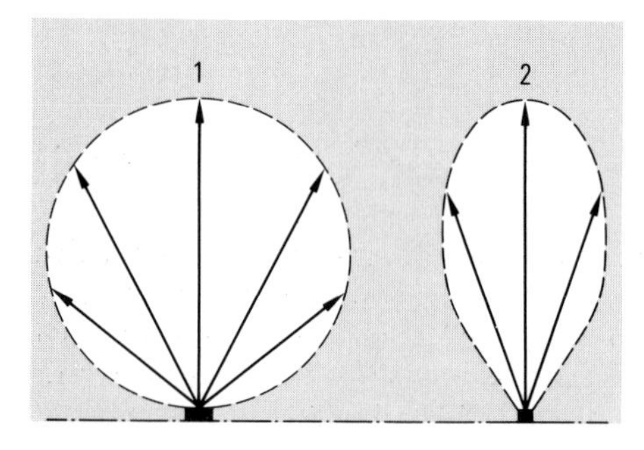

Figure 12.8 Spatial distribution of the emission of (1) an LED and (2) a laser diode

Structure and Characteristics

Laser diodes can be assigned to two families according to their structures, characterized by the type of lateral confinement of the waveguide in the laser. From each family, one laser diode will be described as an example of the many types.

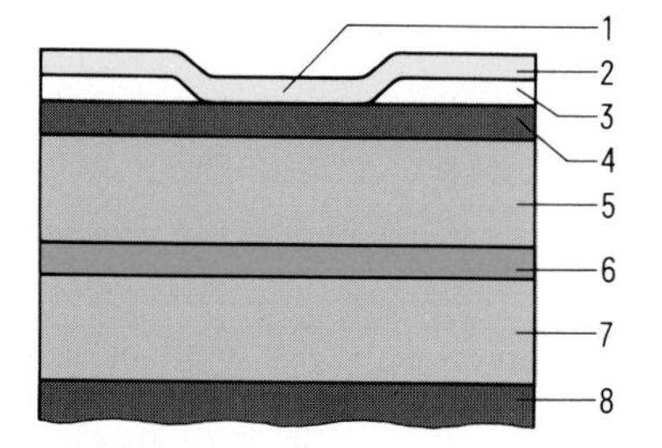

1 Contact window
2 Metallization (TiPtAu)
3 Oxide mask (Al_2O_3)
4 Capping layer (p-GaAs)
5 Confining layer (p-GaAlAs)
6 Active layer (GaAs)
7 Confining layer (n-GaAlAs)
8 Substrate (n-GaAs)

Figure 12.9
Structure of an oxide-stripe laser

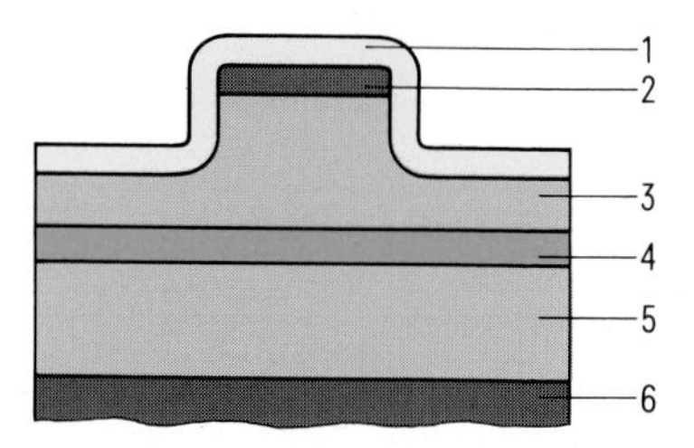

1 Metallization (TiPtAu)
2 Capping layer (InP)
3 Confining layer (InP)
4 Active layer (InGaAsP)
5 Confining layer (n-InP)
6 Substrate (n-InP)

Figure 12.10
Structure of an MCRW laser

Gain Guided Laser Diodes (GLD) (Figure 12.9)

Example

GaAlAs/GaAs oxide-stripe laser for the wavelength range from 800 to 900 nm.

Concentration of the injected carriers in the active strip cause the refractive index to assume a lateral profile, which – in particular, with very narrow stripe widths – provides stable guidance of the laterally limited fundamental mode. This refractive index profile corresponds to a profile of optical amplification – also known as gain. Therefore, the laser diodes of this family are also known as gain guided laser diodes.

Index Guided Laser Diodes (ILD) (Figure 12.10)

Example

GaInAsP/InP-MCRW (metal clad ridge waveguide) laser for the wavelength range from 1300 to 1600 nm.

Laser diodes with built-in waveguides are called index guided laser diodes (ILD) because they have a permanent refractive index profile.

Figure 12.11 shows the influence of temperature on the optical power and the current in a LED and a laser diode.

Table 12.2 lists some typical values and characteristics (at 25 °C) of laser diodes for digital optical transmission.

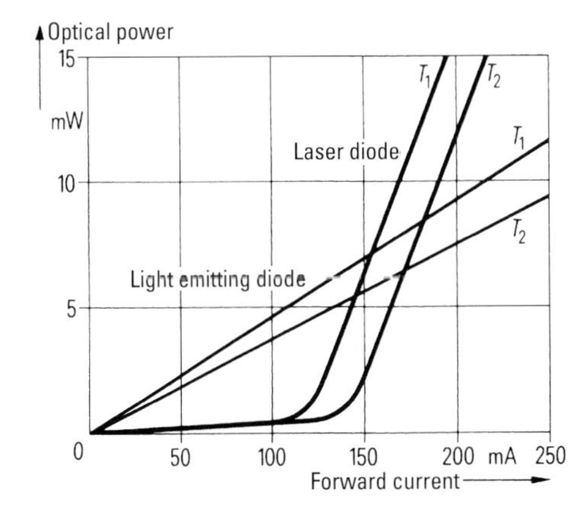

Figure 12.11
Light-current characteristics of a laser diode and light emitting diode at $T_1 = 25°C$ and $T_2 = 60°C$

Table 12.2 Characteristics of laser diodes

			Type	
			DL	DL-MCRW
Wavelength		nm	840	1305
Spectral Width		nm	3.5	3.5
Semiconductor material			GaAl/GaAs	InGaAsP/InP
Structure			Double hetero	Double hetero
Emission			Coherent	Coherent
Light power launched into a fiber with	Core diameter ∅ 50 μm	mW	1	3
	Mode field diameter ∅ 9 μm	mW	–	1.5
Maximum bit rate		Mbit/s	565	1200

DFB Laser

Although laser diodes radiate light in a much narrower wavelength range than LEDs, for special applications in optical data transmission, e.g. for very high bit rates, for coherent transmission or for amplitude modulated television signal transmission, single frequency lasers are needed. For this purpose it is necessary to exactly control the longitudinal (axial) modes of the oscillation of the laser and thereby limit its wavelength spread.

One method for reduction of the number of longitudinal modes is to integrate a periodic structure into the active layer of the laser, which can be coupled dynamically to the radiation field of the laser. In contrast to the Fabry-Perot lasers in general use, which reflect laser light back and forth in a resonator area between two mirror surfaces of the laser, the *d*istributed-*f*eed-*b*ack (DFB) lasers employ a series of wavelike elevations in the semiconductor substrate to reflect only specific wavelength of light and thereby amplify only one resonating wavelength. The spectral width of a DFB laser is smaller than 1 nm (Chapter 10.1.1).

A C^3 laser (*c*leaved-*c*oupled-*c*avity-laser) makes it possible to limit the modes of a laser to only one longitudinal mode. It is composed of two electronically separated Fabry-Perot lasers, which are joined optically by an air space. Only those wavelengths which both resonators have in common will be amplified, so that a highly monochromatic light ray is generated. Technically these C^3 lasers are produced by splitting an ordinary semiconductor laser diode along a crystal surface parallel to the end faces.

VCSEL

The basic design of a laser diode used in the majority of fiber optic communication devices is such that the generated laser light exits at the edge of the laser chip, i.e. the light of these so-called edge-emitting lasers leaves the semiconductor from a lateral surface (edge), perpendicular to the current flow direction.

Lasers have also been developed in which the radiation exits perpendicular to the layer structure (in the current flow direction) through one of the top surfaces. Lasers of this type are called *v*ertical-*c*avity *s*urface-*e*mitting *l*asers (VCSEL). They require mirrors both above and below the active zone, so that a resonator can be formed perpendicular to the layer structure. These lasers were originally developed for transmission around 850 nm. The research is concentrated on shifting the emission wavelength toward 650 nm for plastic optical fibers and toward longer wavelenths of 1300 nm for singlemode fibers and higher bit rates.

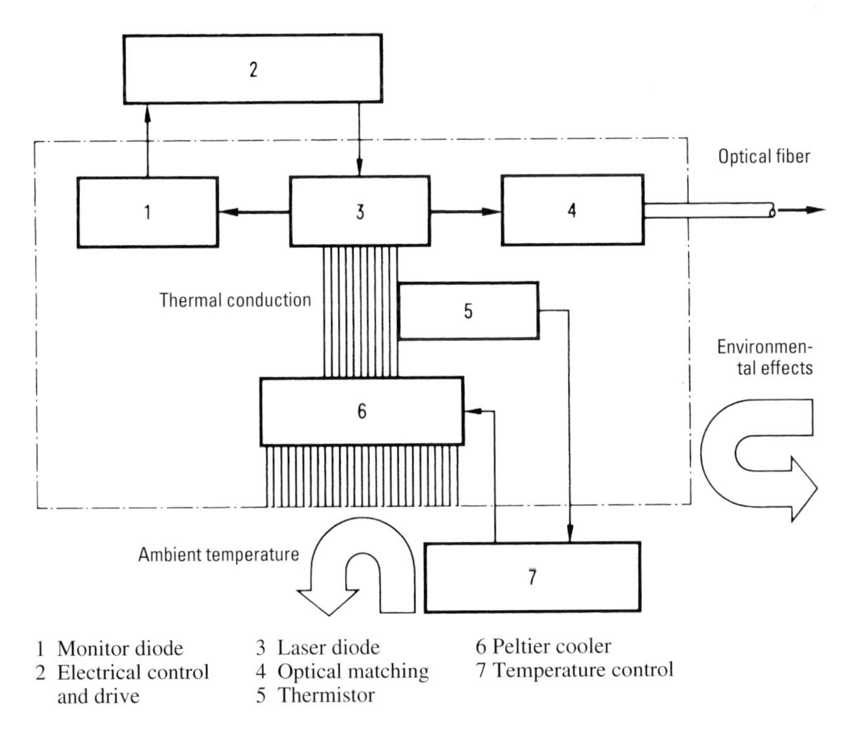

1 Monitor diode
2 Electrical control and drive
3 Laser diode
4 Optical matching
5 Thermistor
6 Peltier cooler
7 Temperature control

Figure 12.12 Schematic of the modular design of a laser module

Laser Module

Practical operation of laser diodes requires several functional groups. These are housed in a hermetically sealed enclosure. The entire unit is called a laser module.

Figure 12.12 gives a schematic of a laser module. The modular design assures that the laser diode can easily be matched to the transmission route.

Optical matching takes place in the module, i.e. short pigtail is optimally coupled to the emitting surface of the diode (fiber alignment is necessary) and is fed out, maintaining the hermetic seal of the module.

The functional group for temperature control is particularly important. Peltier coolers, among other, are used for this purpose.

12.2 Detectors

A photodiode takes advantage of the effect of absorption of light rays in semiconductors. In this effect, irradiation of an energy charged photon with a higher energy level than that of the energy gap E_g sets a pair of electron holes free (reverse of the effect of generating light) and is used to transport electricity.

In theory semiconductor pn junctions can be utilized not only to excite carriers through light injection (Section 12.1.1) but also to detect optically excited charge carriers through separation in the electrical field of the depletion layer, i.e. to receive light. Figure 12.13 provides a schematic of this process. Both the charge carriers generated within the depletion layer of the pn junction and the charge carriers within the "diffusion region" can be detected. Although in the case of the former the two generated charge carriers are immediately separated by the electrical field in the depletion layer, in the latter the minority charge carriers must first be diffused to the pn junction before they can be drawn out by the depletion layer and transported into the opposite neutral semiconductor region. Due to these two processes an electrical current flows in the outer circuit.

For production of such receivers, silicon and germanium or the III-IV-V compounds are most frequently used. Figure 12.14 shows the absorption coefficient α of frequently used semiconductor materials as a function of the wavelength λ.

From Figure 12.14 it can be seen that Si is most suited for the range below 1000 nm. For higher wavelengths, Ge is more advantageous. In the range from 1300 to 1600 nm InGaAs or InGaAsP is well suited as a photodiode material.

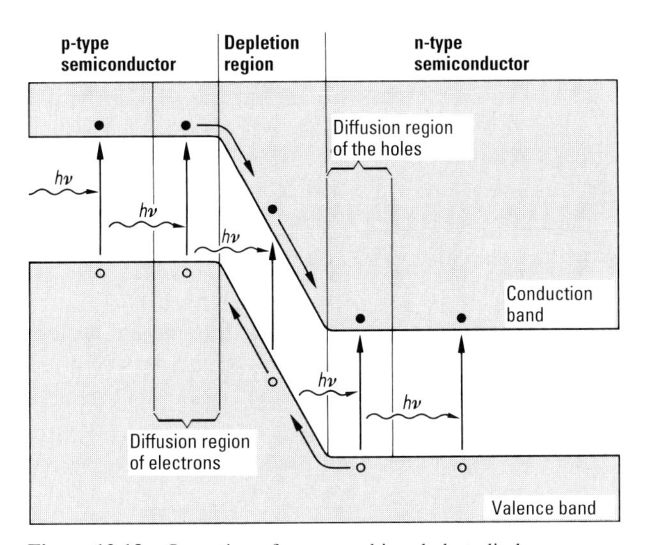

$h\nu$ Photon
● Electron
○ Hole

Figure 12.13 Operation of a reverse-biased photodiode

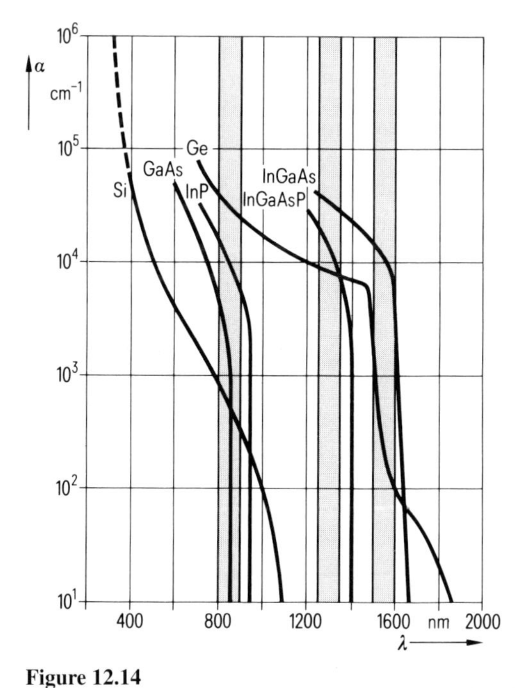

Figure 12.14
Absorption coefficient α as a function of the wavelength λ of semiconductor materials

12.2.1 PIN Photodiode

In semiconductors with low absorption coefficients, the insertion of an undoped semiconductor layer (i-region, intrinsic) between the p- and n-type semiconductor enlarges the region for absorption of radiation. This is a so-called PIN photodiode.

Silicon Photodiodes for Wavelengths up to 1100 nm

Figure 12.15 shows a schematic of the design of a diode with a $p^+ \nu n^+$ structure, where ν indicates that the i-region is n-conductive. Infrared light enters the diode through the p-region, the surface having a Si_3N_4 coating to prevent reflection losses.

*F*ield *e*ffect *t*ransistors (FET) can be used to increase the sensitivity of PIN photodiodes. These hybrid-designed PIN-FET modules can achieve very high sensitivity.

Table 12.3 gives characteristics of photodiodes.

12.2.2 Avalanche Photodiodes

When the acceleration of charge carriers in the electrical field reaches such high speeds that further charge carriers are created by impact ionization, a particularly strong photocurrent is created. This process is called avalanche breakdown and the corresponding photodiode is called an APD (*a*valanche *p*hoto*d*iode).

Silicon Avalanche Photodiode (APD) for Wavelengths up to 1100 nm

Basically, the PIN photodiode described above could also be operated as an avalanche photodiode. The voltage required to achieve the high field strength needed for avalanche breakdown is very high.

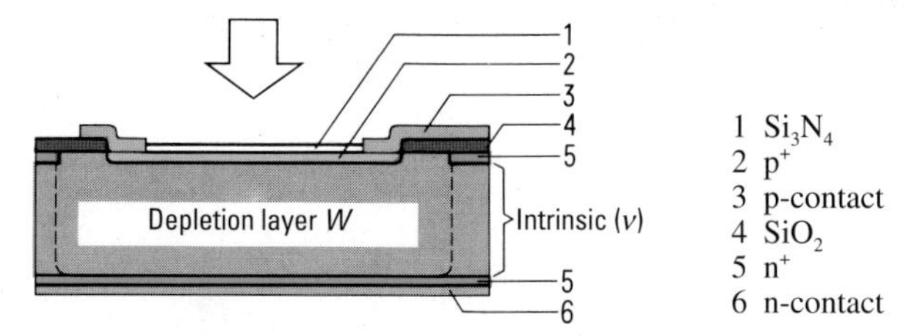

Figure 12.15 Schematic of the design of a PIN photodiode

Table 12.3 Characteristics of receiver diodes

Type	Wavelength (nm)	Semi-conductor material	Structure	Amplification	Spectral sensitivity (A/W)
PIN-PD	500 to 1000	Si	Planar/PIN	–	0.5 to 0.8
APD	500 to 1000	Si	Planar	10 to 100	
PIN-PD	1000 to 1500	Ge	Mesa/planar	–	
APD	1000 to 1500	Ge	Mesa/planar	10	
PIN-PD	1000 to 1600	InGaAs/InP	Mesa/planar	–	
APD	1000 to 1600	InGaAs/InP	Mesa/planar	10 to 50	

The design of an APD is shown in Figure 12.16. For technical reasons such diodes are built only in a p^+pp design.

Figure 12.17 shows a comparison of the sensitivity of receivers with a silicon PIN photodiode and an APD at a bit error rate of 10^{-9}.

InGaAs/InP Avalanche Photodiodes for 1300 nm

A schematic of the design of this diode is shown in Figure 12.18. Just as with the PIN diode, illumination takes place through the substrate; however, the pn junction is not located in the absorbing InGaAs layer but in the InP layer.

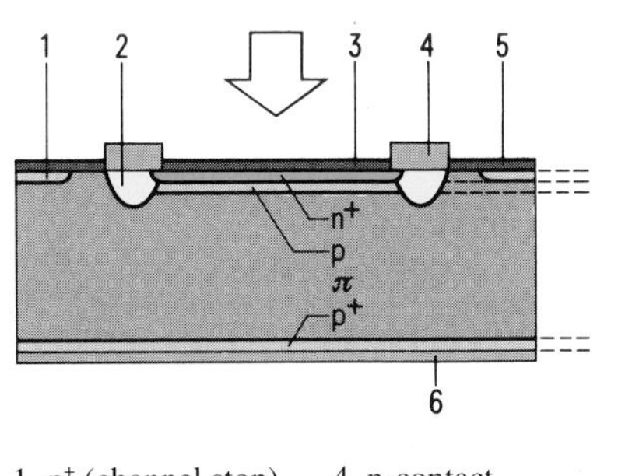

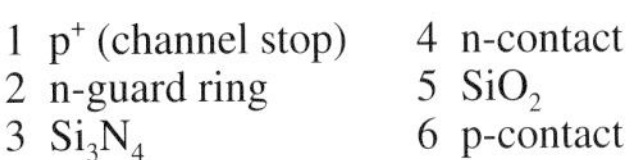

Figure 12.16
Schematic of the design of a silicon avalanche photodiode

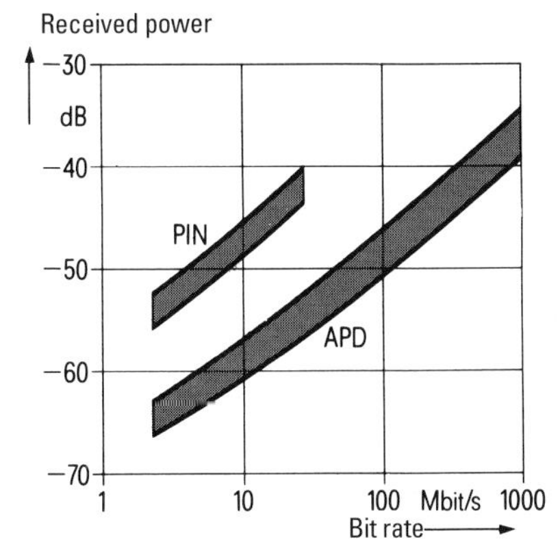

Figure 12.17
Sensitivity of a PIN photodiode and an APD

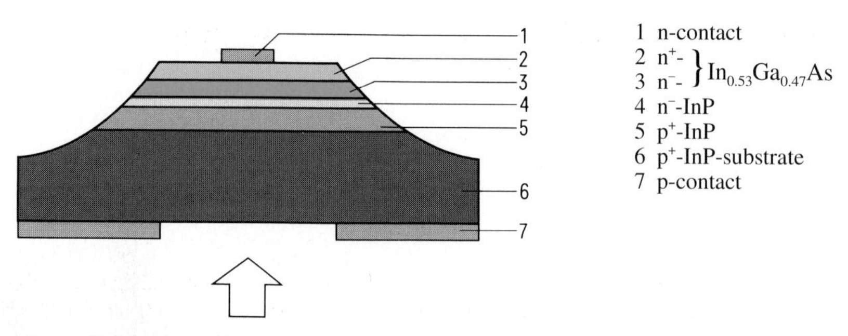

Figure 12.18 Schematic of the design of an InGaAs/InP avalanche photodiode

As a result, a region of high electrical field strength is formed in the n^--InP layer, where avalanche breakdown can be achieved more readily. In order to collect holes generated by photons, it is only necessary to establish a low field strength in the InGaAs layer.

Other diodes for the wavelength range at 1300 nm are:

▷ germanium avalanche photodiodes, an APD that is relatively easy to manufacture,

▷ the combination of an InGaAs/InP photodiode with an GaAs-FET.

Table 12.3 gives characteristics of APDs.

13 Fiber Optic Modules and Components

In addition to fiber optic cables, fiber optic components and modules are of importance in optical transmission systems to amplify, to compensate for dispersion effects, and as routing elements to join or separate information. The information is represented in the form of optical transmission signals, typically called channels. Terminating or connecting elements are connectors (Chapter 10.3.4).

The main limiting factors in today's (single-mode) fiber optic transmission systems are attenuation, dispersion, and non-linear effects such as four-wave mixing, stimulated Brillouin scattering, and self- and cross-phase modulation. Over the last decade, modules have been developed to overcome the attenuation and the dispersion by optical means, without the need to convert the optical signal into the electric domain. The attenuation caused by the fiber itself and the insertion loss of the components and connecting elements can be compensated by direct amplification of optical signals using rare earth doped silica fibers.

Chromatic dispersion compensating modules based on single-mode fibers with a highly negative chromatic dispersion are used to compensate the pulse width widening caused by the single-mode fiber (measured in ps/(nm·km)). These two modules expand the link length, defined as the maximum length between two amplification points, of today's long-haul optical networks to approximately 100 km depending on channel bit rate and channel counts. In some submarine applications link lenghts of more than 300 km have been achieved.

Strong efforts are currently being directed to overcome Polarisation Mode Dispersion (PMD). PMD is receiving much attention because of the increasing channel bit rates to be transmitted on existing optical fiber links which may have a higher polarisation dispersion than optical fibers recently introduced onto the market.

Optical couplers are needed when information is to be transmitted in both directions via one optical fiber or if several channels are to be united in one optical network structure. Couplers are multiport devices, i.e. they have at least three optical ports through which light can be coupled in and out. They are categorized according to their application as wavelength-independent and wavelength-selective couplers. Active routing of optical transmission channels can be performed with optical switches.

13.1 Optical Fiber Amplifiers

A method has been developed for direct amplification of optical signals, especially in the wavelength range of about 1550 nm, by using Erbium-doped optical fibers. Erbium, element 68 in the periodic table of the elements, is a lanthanide in the group of the rare earth elements (Table 3.1). An optical fiber 10 m long with an Erbium-doped core is already sufficient to serve as an EDFA (*E*rbium-*d*oped *F*iber *A*mplifier).

It functions as follows:

The optical signal to be amplified (at a wavelength of 1550 nm) enters a wavelength of 1550 nm into a wavelength-selective coupler (Chapter 13.4) and pump light at a wavelength of approximately $\lambda_p = 980$ nm or 1480 nm is also coupled in.

Figure 13.1 shows a typical configuration with the copropagation of the signal and the pump. It is also possible that the signal and the pump light counterpropagate. In both cases, the signal and the pump light are superimposed and enter the Erbium-doped fiber in which the transmission signal is amplified by induced (stimulated) emission of photons. In the configuration shown in Figure 13.1, an optical isolator is inserted after the Erbium-doped fiber, in order to suppress oscillations from optical feedback. A power split-

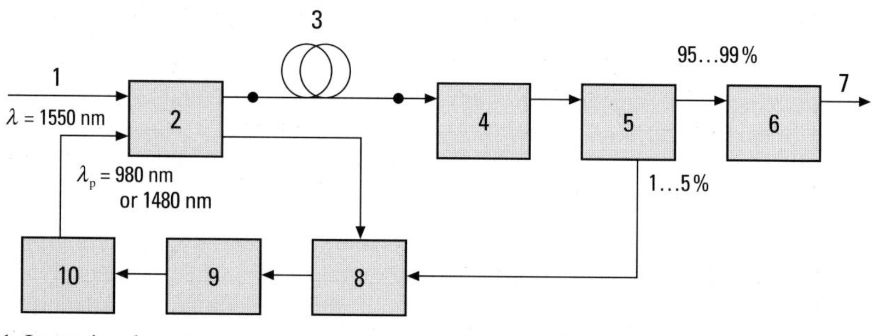

1 Input signal
2 Wavelength selective coupler
3 Erbium-doped fiber
4 Optical isolator
5 Wavelength independent coupler
6 Bandpass filter
7 Amplified output signal
8 Amplifier control
9 Pump laser current
10 Pump laser

Figure 13.1 Schematic of an Erbium-doped fiber amplifier

ter (e.g. 95% of the light is directed to the output and 5% is backcoupled) sends part of the signal to the amplifier controller which serves to regulate the amplification by controlling the pump laser current. A filter prevents pump light as well as noise from spontaneous emission from entering the outgoing optical fiber.

Optical amplifiers are classified according to their application for digital or analog signal transmission and their location in the system: Booster amplifiers are placed after the transmitter to increase the power level to its maximum before coupling the signals into the transmitting fiber. Boosters have a high saturation output power. In-line amplifiers, which are located between two transmission links without conversion into the electrical domain, provide high amplification. Pre-amplifiers have a low noise figure and a moderate amplification and are placed in front of receivers.

One of the biggest advantages of optical amplifiers as compared to semiconductor amplifiers is their use in multichannel applications. Simultaneous amplification of more than 64 channels has been successfully demonstrated. Recent efforts have broadened the amplification spectrum of Erbium-doped amplifiers. They can be used in the C-band (approx. 1525 nm to 1565 nm) as well as in the L-band (1565 nm to 1625 nm).

Amplification of up to 40 dB is possible. In addition to low coupling loss, the linear response of the amplifier is particularly advantageous. Therefore, the optical fiber amplifier is suitable for analog signals, where for example in the distribution of analog TV signals an improvement of approx. 10 dB can be achieved in comparison with a system without optical amplification. This translates to a doubling of the range or a 10:1 increase in the splitting ratio, meaning that approximately ten times as many subscribers can be served.

Even greater advantages can be gained for digital signal transmission, in which the entire optical amplification of approx. 20 dB can be utilized. Furthermore it is possible to cascade (connect back to back) optical amplification route segments. Further characteristics are high output power, low crosstalk, and amplification independent of polarization.

13.2 Dispersion Compensating Modules

Basically two dispersive effects limit the transmission link lengths of todays systems: Chromatic dispersion (Chapter 5.4) and polarization mode dispersion (Chapter 5.10).

Both effects broaden the pulse widths of the transmitted signals which results in overlapping bits, a decrease of the eye opening, and an increase in the bit error rate. The main difference between these two dispersive effects is their statistical behavior in a cable plant: Chromatic dispersion is a linear and deterministic effect whereas polarization mode dispersion is a time dependent and stochastic effect.

(Chromatic) Dispersion Compensating Module (DCM)

Since the invention of the Erbium-doped silica fiber, chromatic dispersion has become the main limiting factor in optical transmission. Dispersion Compensating Modules allow for the compression of broadened optical pulses in the time domain. In a standard single-mode fiber (e.g. ITUT-T G.652) the chromatic dispersion at a wavelength of 1550 nm is approximately 17 ps/(nm · km); i.e. the spectral parts of the light at longer wavelengths travel more slowly than those at shorter wavelengths.

The basic principle for dispersion compensation can be described as a frequency conversion: Spectral parts, which had traveled along the transmission fiber more slowly than others are transmitted through the DCM at a higher speed and vice versa. Today two main techniques are available to realize a DCM:

▷ Fiber Bragg Grating (FBG)

▷ Dispersion Compensating Fiber (DCF).

In FBGs the spectral parts of the signal (wavelengths) are reflected at different locations of a chirped FBG and therefore experience different link lengths. As illustrated in Figure 13.2, the spectral parts at shorter wavelengths are reflected later than those parts at the long wavelengths. The output coupling of the signals is done via an optical circulator acting as a directional, three port coupler.

DCFs have a high negative chromatic dispersion in the wavelength range between 1525 nm and 1625 nm. Complete compensation can be achieved if

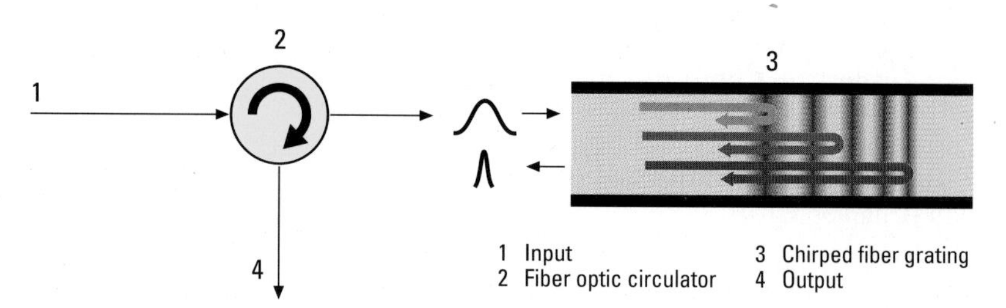

Figure 13.2 Principle of chromatic dispersion compensation with Fiber Bragg Gratings

the absolute values of the dispersion-fiber length-products of the transmission link and of the DCF are equal:

$$M_{SMF} \cdot L_{SMF} = |M_{DCF}| \cdot L_{DCF}$$

M_{SMF} Chromatic dispersion of the transmission fiber

L_{SMF} Length of the transmission link

M_{DCF} Chromatic dispersion of the dispersion compensating fiber

L_{DCF} Length of the dispersion compensating fiber

Since the chromatic dispersion of the fibers is wavelength dependent, so-called slope compensation is necessary for effective broadband or multi-channel application. Slope compensation means that the chromatic dispersion of the DCF not only has a different sign and a much larger absolute value than that of transmission fiber, but also a different sign and approximately the same absolute value of its chromatic dispersion slope (first derivative). Figure 13.3 shows the chromatic dispersion spectra of a standard single-mode fiber as well as the respective dispersion compensation fibers with and without slope compensation.

Although FBGs exhibit a smaller insertion loss, DCF based modules are preferred because they show a higher robustness, are temperature insensitive, and have a better uniformity compared to FBGs. The biggest advantage is the broadband application of slope compensated DCFs. The newest developments in this field are trimmable or tunable DCMs.

PMD Compensator (PMD-C)

The Polarization Mode Dispersion PMD of cable plants depends on many factors. It is a stochastic parameter varying with time. The reason for PMD in fiber optic waveguides is that the two transmitted polarization modes experience different group delay due to mechanical stresses and asymmetries in the waveguides or at connecting points. These two polarization modes (often called principle states of polarization) are the Transverse Electric (TE) and the Transverse Magnetic (TM) waves. A typical maximum PMD value of a standard single-mode fiber is $0.5\ ps/\sqrt{km}$.

The PMD is of special importance for ultra-high speed transmission rates of 10 Gbit/s per channel and more. In cable plants with old fibers of higher PMD values it may become significant even at lower channel bit rates.

Since PMD is a stochastic parameter it is not as easy to compensate for as for chromatic dispersion. If PMD compensation is necessary this has to be done dynamically, though no PMD compensators are currently commercially

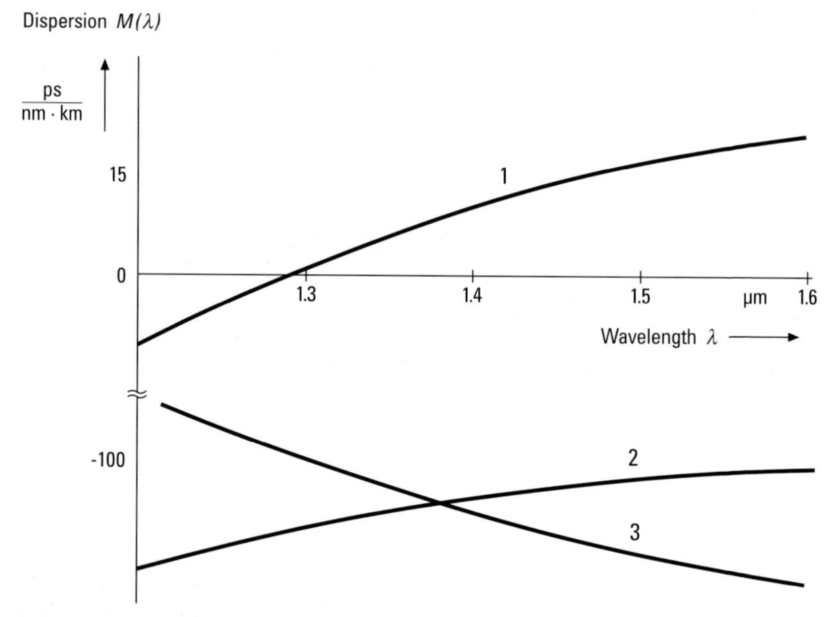

1 Standard single-mode fiber (ITU-T G.652)
2 Dispersion compensating fiber for G.652
3 Dispersion compensating fiber for G.652 with slope compensation

Figure 13.3 Schematic chromatic dispersion spectra of a single-mode fibers

available. However, various research and development groups are working on solutions with different approaches and have shown results at various conferences. The common, basic principle of a PMD-C is shown schematically in Figure 13.4 and can be described as follows:

The optical signal is separated into its two polarization states. Each polarization state is delayed by one or more variable optical delay lines (e.g. Lithium Niobate crystals) by a control mechanism. This control mechanism receives information about the quality of the signal (e.g. BER or eye opening) from a monitor which receives a small portion of the signal.

The realization is difficult because the controller has to measure and process the required information (e.g. eye opening) at high data rates and faster than the time constants of the PMD variations. Depending on where the cable plant is installed, PMD variations over time can be as high as a few ps.

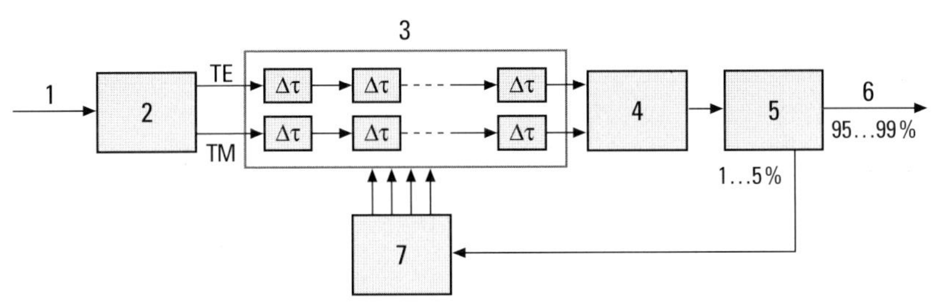

1 Input signal (TE and TM)
2 Input polarization beam splitter
3 Variable optical delay lines
4 Output polarization beam splitter
5 Wavelength independent coupler
6 Output signal (TE and TM)
7 Controller

Figure 13.4 Schematic drawing of a PMD compensator

13.3 Wavelength-Independent Couplers

$M \times N$ couplers are used to optically connect M optical waveguides with N different waveguides at one point in a fiber optic transmission network (Figure 13.5). These couplers offer nearly wavelength-independent connection of the optical fibers within a specified wavelength range.

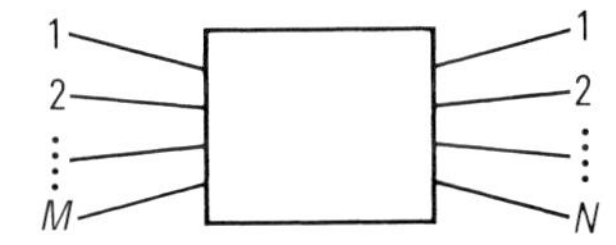

Figure 13.5 $M \times N$ coupler

In addition to these general $M \times N$ couplers the following special designs are being used:

$M = N$

This arrangement with an equal number of input and output ports is called a star coupler or mixer, e.g. 4 × 4 star coupler (Figure 13.6).

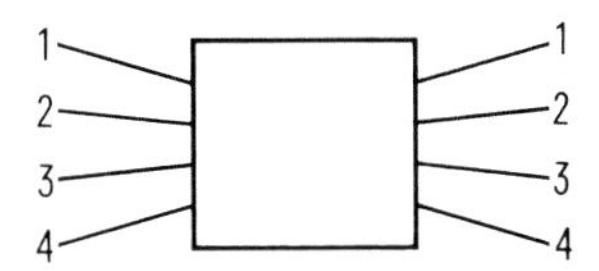

Figure 13.6 4 × 4 star coupler

M = 1, N unspecified

This configuration with only one input and several output ports is known as splitter, tree, or branching coupler, e.g. 1 × 4 splitter (Figure 13.7).

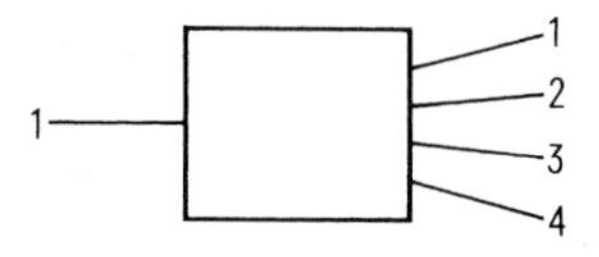

Figure 13.7 1 × 4 splitter

M unspecified, N = 1

This configuration is the mirror image of the previous configuration. Several input ports are connected to one output port. This is called combiner, e.g. 4 × 1 combiner (Figure 13.8).

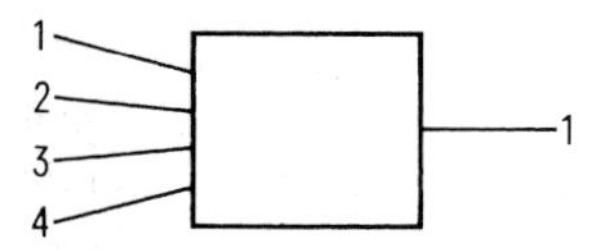

Figure 13.8 4 x 1 combiner

With regards to the techniques for manufacturing fiber optic couplers, a differentiation is made between fused-tapered fiber couplers and planar waveguide couplers.

The fused-tapered fiber coupler is most widely used. To manufacture them, optical fibers are partially melted and drawn a few millimeters. The mode field of one optical fiber thereby overlaps that of the other and light is coupled.

The planar technique offers a different possibility for coupling. It is achieved through use of one of several different technologies, whereby a refractive structure is created which is suitable for light guidance in a waveguide. Examples of these technologies are flame hydrolysis deposition, (plasma enhanced) chemical vapor deposition, and ion exchange in glass. Planar couplers and splitters have the advantage of being highly reproducible and have the potential for mass fabrication.

Besides the core or front end technology – fused-tapered fibers or planar waveguides – the packaging or back end technology becomes more and more important when the devices are installed in harsh environments, strongly influencing the reliability of the devices. Recent developments for fused-tapered couplers show significant improvement of their reliability (e.g. Multiclad® technology).

13.4 Wavelength-Selective Couplers

The tremendous demand for optical bandwidth in today's networks can only be satisfied with multiplexing technologies. In parallel to the rapid installation of fiber optic cables (Space Division Multiplexing, SDM) and the tendency to ultra-high data rates (Time Division Multiplexing, TDM), the application of Wavelength Division Multiplexing (WDM) systems is increasing today. Figure 13.9 shows an overview of the three different multiplexing technologies applied in today's optical fiber networks.

Space Division Multiplexing

With optical fiber multiplexing each signal is assigned to one optical fiber and one wavelength. The routing and management is relatively easy but space consuming.

Time Division Multiplexing

Several signals are interleaved electrically to give one composite signal which is transmitted via an optical fiber at one wavelength after electrooptic conversion. This requires a highly complex electronic system when it comes to high data rates to separate individual lower data rate channels.

Wavelength Division Multiplexing

In a WDM system, electrooptic transducers operate at different signal center wavelengths. Their optical power is combined via a wavelength-selective coupler into one optical fiber. Depending on its application of combining or separating signals, they are called wavelength division multiplexers or wavelength division demultiplexers, respectively. When many signal wavelengths are multiplexed (e.g. more than 4) or the spacing between these signal wavelengths is small (e.g. smaller than 4 nm) it is often called Narrow Bandwidth (NBWDM) or Dense (DWDM) Wavelength Division Multiplexing.

There are one- and two-directional WDM couplers. With a one-directional WDM coupler, signals are transmitted via one optical fiber at different wavelengths in the same direction; with a two-directional WDM coupler, they are also transmitted in the opposite direction. Wavelength division multiplexers/ demultiplexers each contain one continuous optical fiber.

The difference between the signal wavelengths – often called channel spacing – lies at approximately 200 nm if light emitting diodes are used as transmitters. In systems with laser diodes the channel spacing is typically at or below 200 GHz. The channel frequency grid is standardized to ITU-T recommendation for long-haul applications. There are different techniques used to manufacture WDM couplers.

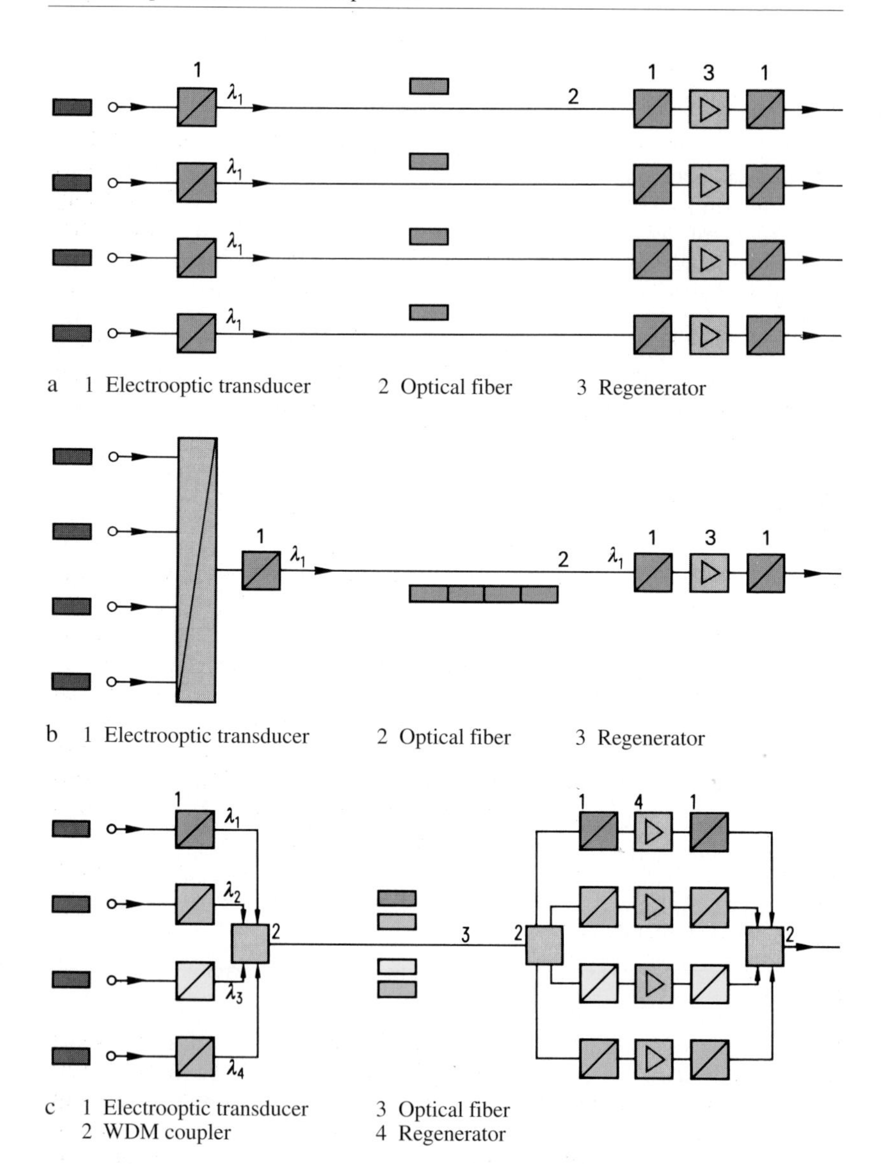

Figure 13.9 Overview of various multiplexing technologies in optical fiber networks

a Space division multiplex
b Time division mulitplex
c Wavelength division multiplex

Fused-Tapered couplers

If only two channels are to be multiplexed, the Fused Biconic Tapered (FBT) coupler is applicable. Two optical fibers are coupled with each other over a certain distance called the coupling zone. This coupling is done by aligning the two fibers in parallel melting, and drawing until the required optical transmission performance is achieved. The main application for FBT is to combine the signal and pump power for Erbium-doped fibers. They are also used as wavelength-selective directional couplers in CATV systems (1310 nm and 1550 nm) or to couple supervisions signals e.g. at 1625 nm wavelength into a 1550 nm wavelength signal system. The main disadvantage of FBTs is that they work efficiently only for 2 channels. DWDM applications are therefore not possible.

Fiber Bragg Grating Filters

A wavelength selective, reflection type filter can be achieved by a Fiber Bragg Grating (FBG) in the core of a special single-mode fiber. The FBG is written into the core by means of interfering and focusing UV-light. The coupling of the light into the FBG and the filtered wavelength out of the FBG is accomplishing using a fiber optic circulator. The circulator acts as a three port, directional coupler. If more than one wavelength has to be multiplexed or demultiplexed, the FBGs and circulators have to be cascaded as shown in Figure 13.10. This cascading means that the insertion loss of the multiplexer/demultiplexer is strongly dependent on the channel count. The main advantage of FBGs is that the reflection spectrum can be very narrow and would allow narrow channel spacings. However, the reflection spectrum of FBGs typically show ripples which cause phase errors, especially in high speed networks. In addition, it is mechanically difficult to compensate for the temperature dependence of FBGs.

Thin Film Filters

The basic element of Thin Film Filters (TFF) is a bulk glass on which some layers of reflective optical materials are deposited. In contrast to FBGs, here the filters work as transmitting filters (band pass): The required wavelength band to be coupled in or out is transmitted through the TFF whereas the rest of the spectrum is reflected. Therefore a circulator to couple the signal channels in or out is obsolete. State-of-the-art layer deposition techniques allow channel spacings for TFF of 100 GHz and below with excellent cross-talk characteristics. Automated assembly processes show that TFFs have a great potential for mass fabrication. TFFs also show a very low insertion loss temperature dependence and the center wavelengths of the channels. However,

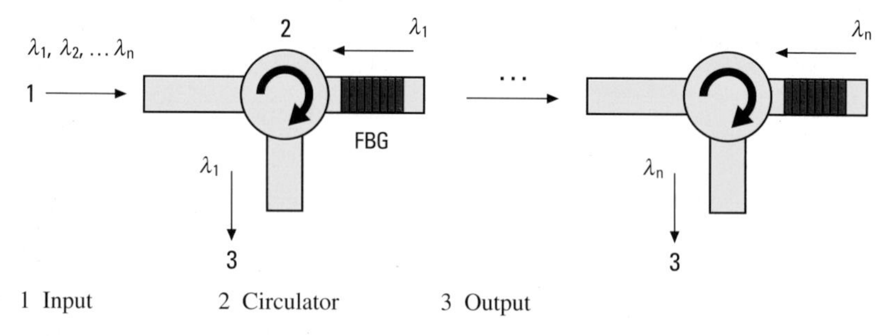

1 Input 2 Circulator 3 Output

Figure 13.10 Schematic drawing of a FBG type demultiplexer

similar to the FBG filters they are a cascaded system as shown in Figure 13.11, resulting in higher insertion losses at higher channel counts. Both, FBG and TFF devices are not bidirectional: They cannot be used as multiplexers and demultiplexers with the same arrangement.

Arrayed Waveguide Grating Filters

A multichannel, wavelength-selective filter in planar lightwave circuit technology is realized with the Arrayed Waveguide Grating (AWG) filter. Another commonly used expression for this device is the Phased Array or

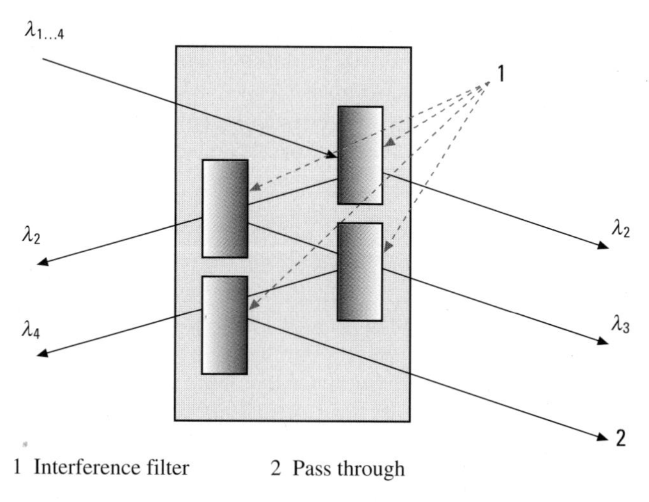

1 Interference filter 2 Pass through

Figure 13.11 Schematic drawing of a TFF type demultiplexer

Phasar filter. The basic principle is similar to an optical monochromator and is explained by Figure 13.12. The signal channels coupled into the first slab waveguide are uniformly distributed and coupled into the phased array. The waveguides of the phased array act as a phase shifter because they have slightly different optical path lengths. When the light at the end of the second slab waveguide interferes, the different wavelengths of the different channels show their intensity maxima at different locations. Using the output waveguide array, the different channels can be coupled into different optical fibers.

Since the AWG multiplexer/demultiplexer is – in contrast to FBG and TFF based devices – a parallel filter, the insertion loss is basically independent of the channel count. Another advantage is that AWGs are bidirectional and can be used as multiplexers and demultiplexers in the same configuration. Besides the relatively high expense for the fiber-chip coupling, the main challenge is the compensation of the temperature dependence of the AWG filter. This dependence results in a shift of the filter center wavelengths. Most commercially available devices are actively thermally controlled: To avoid the necessity of heaters and coolers the AWG chip is heated to a temperature above the operating range, e.g. +75°C. This entails high power consumption and the necessity to control this feature. It also adds an additional possible point of failure as additional heat is generated in the system and thus the ageing of the device is accelerated. However, there is currently one commercially available technique which allows the passive athermalization

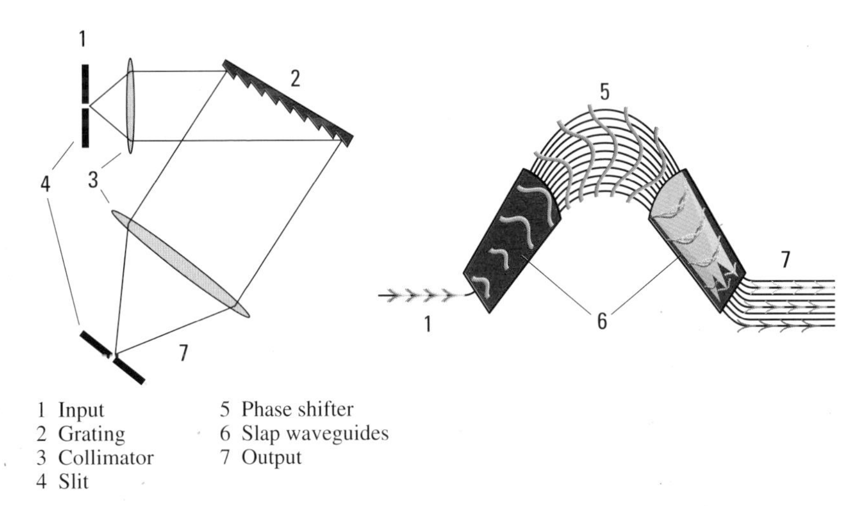

Figure 13.12 Comparison of an optical monochromator (left) with an AWG filter (right)

of AWG filters. This athermalization is achieved by an extension bar which moves the input location into the slab waveguide with temperature. This compensates for the wavelength shift typically induced by the temperature dependence of the AWG chip (Figure 13.13). Corning has shown that this technique can work very reliably with excellent optical performance.

Comparison of DWDM filter techniques

Since FBG and TFF are cascaded filter systems, they are limited in channel count by insertion loss and its variation. FBGs need fiber optic circulators which add some insertion loss and cost. Additionally their insertion loss spectrum shows a ripple which results in signal phase shifts and limits their applicability to lower bit rate systems (e.g. < 2.5 Gb/s). However, the pass-band width of FBGs can be very small so that they might be used when ultra-narrow channel spacings of below 25 GHz are necessary. The latest TFFs (e.g. from Corning) show excellent insertion loss and channel isolation parameters for channel counts of up to 16 for 100 GHz and 200 GHz channel spacing. They are temperature insensitive and the machinery required to automate assembly is available. When it comes to higher channel counts and lower channel spacings, AWGs show significantly better insertion loss and variation thereof. AWGs should be used for channel counts starting at 16 and channel spacings of 100 GHz and 200 GHz. Corning's passive athermalization provides a clear advantage for transmission system equipment suppliers.

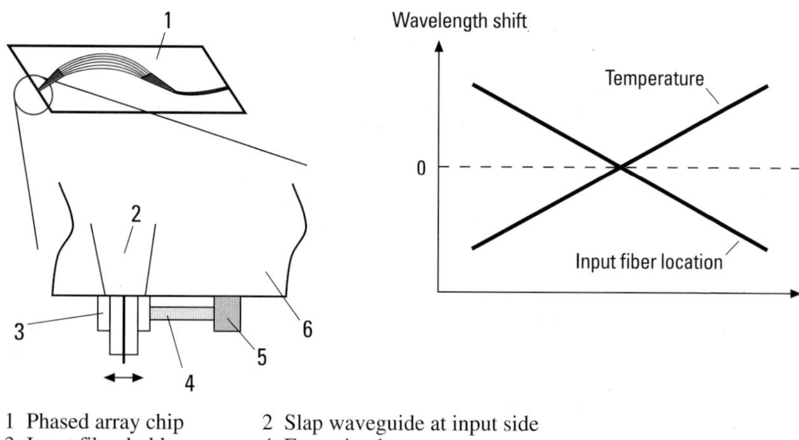

1 Phased array chip
2 Slap waveguide at input side
3 Input fiber holder
4 Extension bar
5 Post for fixing
6 Chip

Figure 13.13 Principle of passive athermalization for AWG filters

Besides FBG, TFF and AWG there are some techniques (e.g. cascaded fused-fiber based MZ interferometers, bulk glass gratings) which have either uniquely niche applications, reliability problems or very low yields during manufacturing.

Add-Drop Multiplexers

In some networks it is necessary to couple out (drop) or in (add) one or more wavelengths from or into transmission fiber and transmit all others. This can be achieved by using special wavelength-selective couplers or wavelength division multiplexers called Add-Drop Multiplexers (ADM). Figure 13.14 shows a schematic of the functional principle of the ADM. The latest development efforts here include configurable ADMs where the operator can select the dropped or added wavelength. This would allow flexible and remote traffic routing in the central office.

13.5 Optical Attenuators

Optical attenuator may sound strange as attenuation is one of the limiting factors in optical transmission systems and special modules (EDFA, Chapter 13.1)) have been developed to overcome it. However, in some applications it is necessary to adjust power levels to certain values. For example, one may

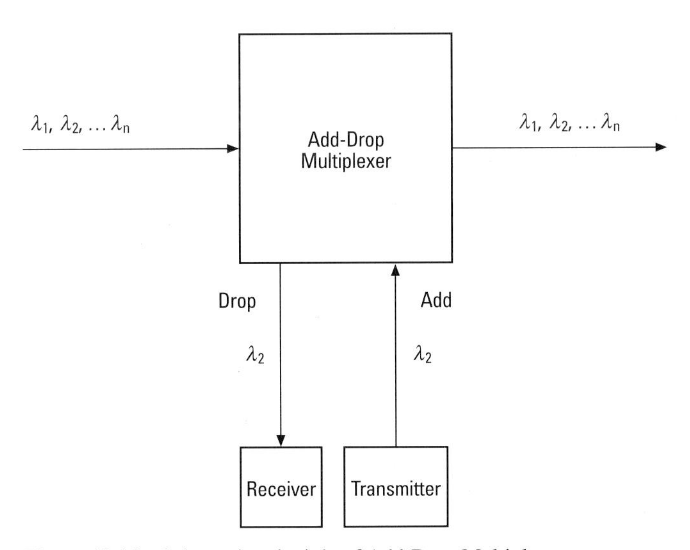

Figure 13.14 Schematic principle of Add-Drop Multiplexers

need to avoid a situation where a module would not work as expected because of saturation by high power levels. Another example is when one channel may influence other channels in a DWDM system because of non-linear effects in the transmission fiber. The power levels of the different signal channels often vary significantly from each other. The reason for this can be variations in the transmitter output power levels, different transmission lengths or the wavelength dependence of the optical components and modules.

In principle there are two different types of optical attenuators: Devices with a fixed or an adjustable value of attenuation. Fixed optical attenuators are realized either by an air gap, a doped fiber or filter glass incorporated into the transmission path as absorption filters, or by bending a fiber in the transmission path. Doped fiber or filter glass-based fixed attenuators are integrated in adapter housings for various connector styles (e.g. SC, FC, and ST). Fixed attenuators are typically implemented as in-line attenuators directly behind transmitters or amplifiers.

Since the performances of optical components and modules vary with time, temperature, etc., a dynamic or variable optical attenuation is necessary. One method to adjust optical attenuation dynamically is to use a mechanism to bend a multiclad coupler. Here, Corning has proven that their technology is highly reliable.

Another approach is to make use of the thermo-optic effect with Mach-Zehnder (MZ) interferometers in planar lightwave circuit technology. Figure 13.15 shows the basic principle of this method where the temperature dependence of the refractive index of the glass is used. The electrodes heat up the waveguide and lengthen the optical of one arm of the interferometer. Depending on the induced phase shift the interference at the output of the MZ interferometer results in a specific attenuation spectrum. For control of the heating current, a small part of the light is coupled out and measured by a detector.

The implementation of planar lightwave circuits also allows the integration of several of the variable optical attenuators on one chip in parallel. This allows a reduction in the size and the number of devices which is very advantageous in DWDM systems: The power levels of several DWDM channels (e.g. 8 or 16) can be adjusted dynamically in one device.

13.6 Optical Switches

Optical switching to route transmission signals is of great importance to avoid the otherwise necessary optical-electrical conversion, especially at high data rates. In general, optomechanical, optoelectronic, thermooptic and

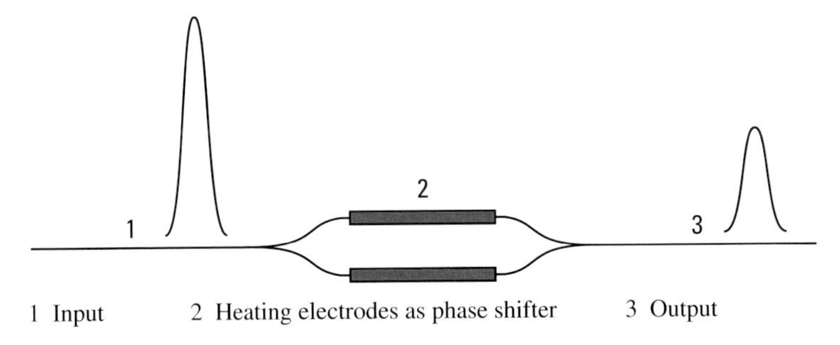

1 Input 2 Heating electrodes as phase shifter 3 Output

Figure 13.15
Schematic drawing of a Mach-Zehnder interferometer in planar lightwave circuit technology

microelectromechanical switches are used today. Besides the number of channels which can be managed, key requirements for optical switches include:

▷ Short switch times (< 20 ms)

▷ Broadband (1525 nm to 1610 nm)

▷ Low cross-talk

▷ Non-blocking architecture in case of power failure

▷ Bidirectional

▷ Remotely (re-)configurable

▷ Low operating voltage, power consumption, and heat dissipation

Optomechanical switches use precision mechanics to switch the position of the input fiber to match with different output fibers. Piezoelements or stepping motors are typically used for this movement. Their main advantage is that the switching mechanism is entirely wavelength and mostly temperature independent. However, switching times are relatively slow (approximately 20 to 500 ms).

Optoelectronic and thermooptic switches are based on planar lightwave circuit technology. Depending on the waveguide material the electrooptic effect (with $LiNbO_3$) or the thermooptic effect (with SiO_2) is applied. Similar to the variable optic attenuator in planar lightwave technology, light can be coupled into different waveguides, depending on the adjusted phase shift when a Mach Zehnder interferometer is used (Figure 13.16). Recent developments with polymer materials show promising results. Optoelectronic and thermooptic switches show fast switching times (< 10 ms) which are necessary

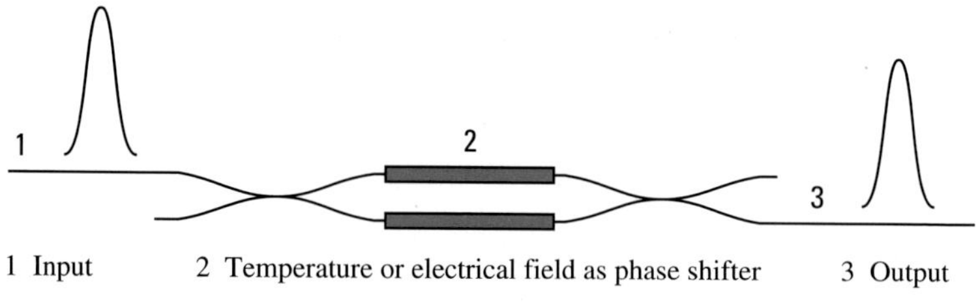

Figure 13.16 Basic principle of a thermooptic switch

in today's optical transmission networks. They also have the potential for a higher level of integration by cascading the Mach-Zehnder interferometer structures, allowing larger switching matrices of 256 × 256 or even higher.

MicroElectroMechanical Systems (MEMS) integrate different technologies (e.g. mechanical, electrical, thermal) to achieve a specific effect. They can be used for different applications in optical components whereof the switching is the most promising one. Here a small mirror slides between different positions to reflect the signal into the desired direction. MEMS are manufactured by using integrated circuit technology. Optical fibers are coupled to the MEMS as input and output interfaces.

14 Systems for Optical Transmission

Optical transmission systems are used for transmission of electrical signals via an optical fiber. Their components are an electrooptic transducer as the light transmitter at the beginning of the route, the actual fiber optic transmission route and the optoelectric transducer as the light receiver at the end of the route. As in systems with metallic conductors, there are line-terminating units at the beginning and end of the route; the units located between them are repeaters for analog transmission and regenerative repeaters for digital transmission (Figure 14.1). Readers who want to learn more about this very broad topic should consult the specialized literature.

Optical and electric transmission systems have the *same* electrical interfaces. This means that one important goal of the introduction of optical fiber technology was achieved that in particular makes integration into existing systems much easier.

As the method of transmission, digital technology has also established itself for optical fibers, because it permits free combination of bit rates from a great variety of sources (telephone-data networks, etc.). Analog signal transmis-

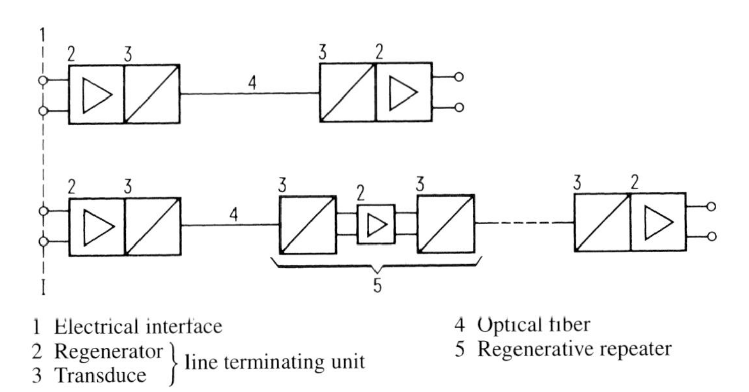

Figure 14.1
Basic components for optical transmission without regenerator (above) and with regenerative repeater (below)

sion continues to decline in importance since the introduction of optical fiber technology and is only used for special applications.

14.1 Digital Transmission Technology

The most important functions of digital transmission technology are digitizing of analog signals (mainly voice), multiplexing of digital signals and transmission of these digital signals, e.g. via optical waveguides. With the introduction of digital technology the conditions became available for integration of the services, such as telephone, telex, data transmission and telefax.

In ITU-T, transmission systems with specific bit rates were established for this purpose (Table 14.1).

▷ PDH (*P*lesiochronous *D*igital *H*ierarchy)
The PCM30 system transmits at the lowest level of this hierarchy. In it 30 telephone channels with 64 kbits/s are combined for a bit rate of 2048 kbit/s. For short they are called systems for transmission of 2 Mbit/s signals. PDH is a standard which has been in effect since the 1960s and is used by the Deutsche Telekom AG and many telecom authorities around the world.

Tabe 14.1 Transmission systems for digital signals

Plesiochronous Digital Hierarchy	
Number of 64 kbit/s channels	Bit rate (rounded) in Mbit/s
30	2
120	8
480	34
1920	140
7680	565[1])
Synchronous Digital Hierarchy	
STM level[2])	Bit rate (rounded) in Mbit/s
STM-1	155
STM-4	622
STM-16	2488

[1]) Not an ITU-T hierarchy level
[2]) Synchronous transport module

▷ SDH (*S*ynchronous *D*igital *H*ierarchy)
The STM-1 level is the basic level of this hierarchy. SDH levels provide the transport capacity for digital signals, e.g. from the PDH hierarchy. SDH is a standard set by ITU-T in 1988 and systems based on it are now being used in the networks. Due to the fact that they can be applied universally, SDH systems (Chapter 14.1.2) will be in frequent use in the growing transmission networks. In the long run they will replace PDH technology.

Note

The bit rate can be described as follows:

The original analog signals can be regained from pulses created by sampling, when the sampling frequency is at least twice as large as the highest signal frequency. Because the telephone voice frequency range goes up to 3400 Hz and additional capacity must be reserved for filters, it was agreed to allow for 4000 Hz and a sampling frequency of 2 × 4000 Hz; this therefore gives 8000 Hz. By encoding each impulse into an 8-bit word, the bit rate is obtained, which is available for each individual telephone signal, namely 8 kHz × 8 bit = 64 kbit/s. The bit rate of PCM30 systems is 64 kbit/s × 32 = 2048 kbit/s.

Systems with a higher number of channels transmit bit rates of 8, 34, 140 or 565 Mbit/s; they are based on the 2 Mbit/s signal of the PCM30 multiplexing unit and are created with digital signal multiplexing units.

The basic framework for SDH is STM-1. It is repeated at intervals of 125 μs and is divided into overhead and payload parts (usable information part) of a total of 2430 octets. For them the bit rate is 2430 × 8 × 8000 bit/s = 155.520 Mbit/s.

The payload part can accept containers on four different levels. These levels correspond to the plesiochronous bit rates of 1.5/2, 6, 34/45 and 140 Mbit/s, in both the North American and European hierarchies (Table 14.2). The overhead part contains information for the internal administration of the STM signal and also provides transmission capacity of network supervision and network control (network management information).

Higher STM levels are defined as integral multiples of the first level.

For all of the PDH and SDH hierarchy levels listed, there are line terminating units suitable for fiber optical transmission.

14.1.1 PDH Systems

Digital signals are combined for transmission in steps of 2, 8, 34, 140 or 565 Mbit/s. For example, at the 2 Mbit/s level up to 30 telephone calls and at the 34 Mbit/s level up to 480 telephone calls can be transmitted by the time division multiplex method. Specific hierarchy levels have been established for both network levels (local and long distance), which have specific characteristics corresponding to the network level.

Local Network

34 Mbit/s and 140 Mbit/s systems are in use in local networks. They are interconnected via optical local intracity cables. Because these are point to point connections with relatively short route lengths of only a few kilometers, systems can be utilized that by comparison with the long distance network are less complicated. For example no regenerative repeaters are needed.

Long Distance Network

For the communication system (trunk network) which is subdivided into regional and national networks, optical fiber systems are being used which are tailored to this network structure, i.e. spanning long route lengths. Therefore, in contrast to the systems for the local network, they have technical dimensions suitable for insertion of regenerative repeaters and the control units they require.

One important precondition for realization of these optical fiber systems was the availability of suitable optical transmitters and receivers. Laser diodes (Section 12.1.2) are often utilized for such applications as optical transmitter modules and PIN–FET diodes (Section 12.2.1) are often used as receiver modules. Their operating wavelengths are 1300 nm. Single-mode optical fibers (9/125 µm) are used for transmission.

In the regional long distance network 34 Mbit/s and 140 Mbit/s are most frequently used. In the national trunk network with very long distances and a very high capacity requirement, 140 Mbit/s or 565 Mbit/s systems are mainly used.

14.1.2 SDH Systems

Since 1990 the Deutsche Telekom AG has been introducing synchronous line equipment with a transmission capacity of 622 Mbit/s (SLA 4) and 2488 Mbit/s (SLA 16) into its network. Initially these are replacing

565 Mbit/s systems in the long distance network. Because SLA 4 and SLA 16 can transmit both PDH and SDH signals, no complications arise when they are integrated into an existing PDH network. As network elements with synchronous interfaces (e.g. multiplexers, cross connectors) come into wider use, this synchronous line equipment can also be employed in the regional and local networks.

14.1.3 Examples of Systems with SLA 4 and SLA 16

As an example for all kinds of optical line equipment the synchronous line equipment SLA 4 and SLA 16 are considered more closely.

This system family offers the following functional units:

Synchronous line multiplexers SLX 1/4 and SLX 1/16, synchronous line regenerators SLR 4 and SLR 16 as well as synchronous branching regenerators SLR 4 D/I and SLR 16 D/I (Figure 14.2).

Optical line interfaces for short or long connections over single-mode optical fibers (in the second or third optical window) are available. The interfaces can be switched independenly from one another for synchronous 155 Mbit/s and plesiochronous 140 Mbit/s signals. The drop-and-insert (D/I) regenerators offer direct access to the 140 or 155 Mbit/s signals inside the transmitted main bit stream.

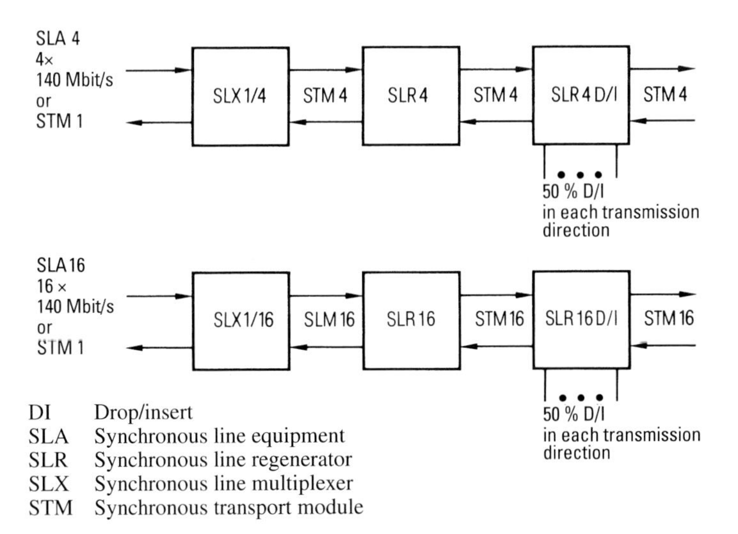

Fig. 14.2 SDH systems SLA 4 and SLA 16

The supervision concept corresponds to ITU-T recommendations and interfaces for local indicators as well as for a telecommunications management network (TMN) are incorporated.

14.1.4 International Trends in Optical Fiber Systems

In Europe, exclusively digital systems have established themselves for transmission with optical fibers. In addition to the usual PDH 2, 8, 34 and 140 Mbit/s systems, new SDH systems are becoming more important. These systems transmit bit rates with 155 Mbit/s (STM-1), 622 Mbit/s (STM-4) and 2488 Mbit/s (STM-16). In addition to transmission technical functions (line terminating units and regenerators), they offer integrated mulitplexing functions for combination of four, six or ten signals of the first synchronous hierarchic bit rate (155 Mbit/s, i.e. STM-1) to create signals at correspondingly higher bit rates by means of a new synchronous, byte oriented multiplexing technique. The systems generally operate in the wavelength range of 1300 nm. Wavelengths around 1550 nm are becoming more important, especially in the trunk network for which achievement of ever longer distances is an important cost factor. Introduction of optical amplifiers (Chapter 13.1) must be seen in this context, because they allow repeater free lengths of several hundred kilometers.

Except in Japan and a few other countries, where slotted core cables (Chapter 9.1) are in use, fiber optic cables with filled loose buffer have established themselves as a transmission medium in public telecommunications networks, whereby single-mode fibers are used exclusively.

Outside of Europe, North American and Japanese developments are important. Just as in Europe the synchronous digital hierarchy has had an important impact on the development of new transmission systems, in the USA it was the creation of a new standard for *s*ynchronous *o*ptical *net*works called SONET. This standard formed the basis for SDH and defined the same bit rates higher than 155 Mbit/s. In the USA other names for corresponding bit

Table 14.2 Comparison of European, North American and Japanese bit rates

	Bit rates (rounded) (Mbit/s)										
Europe	–	2	–	8	–	34	–	–	140	–	565 [1]
United States	1.5	–	6	–	–	–	45	–	135 (90)	405[1]	565 [1]
Japan	1.5	–	6	–	32	–	–	100	–	400[1]	–

[1] In this case there is no ITU-T recommendation

rates appear due to the insertion of a synchronous hierarchic level at 45 Mbit/s (STS-1 and OC-1) (Table 14.2). For example an STS-3 (synchronous transport signal) corresponds to an OC-3 (optical carrier).

Available systems with bit rates of up to 2.5 Gbit/s will be upgraded in capacity to 10 Gbit/s (STM-64).

In the USA optical fibers are cabled in loose buffers as well as in ribbon technology and – for special applications – in slotted core cables.

14.2 Analog Transmission Technology

Fiber optic systems are designed primarily for transmission of digital signals, however there are some suitable applications for analog transmission. This is true e.g. for video transmission in cable television networks and in conjunction with intrusion protection.

When an analog optical fiber system is being selected, particular attention must be paid to the method of modulation. The most important criteria in this case are – in addition to attenuation and usable bandwidth of the optical fiber – the noise levels of the optical transmitter and receiver and the nonlinear distortion due to the kink in the characteristic curve of the transmitter diodes. With direct amplitude modulation of the light, the time curve of the light power is directly proportional to the time curve of the electrical signal. Because good noise and linearity characteristics are technically complex to achieve with this type of modulation, *frequency modulation* is most frequently applied – a method with a signal to noise ratio which is quite insensitive to disturbances of any kind.

15 Appendix

15.1 Abbreviations of National and International Standards Committees

ANSI	American National Standards Institute
CENELEC	Comité Européen de Normalisation Electrotechnique
CEPT	Conférence Européenne des Postes et Télécommunications
DKE	Deutsche Elektrotechnische Kommission im DIN und VDE
DIN	Deutsches Institut für Normung e.V.
ETSI	European Telecommunications Standards Institute
IEC	International Electrotechnical Commission
IEEE	Institute of Electrical and Electronics Engineers
ISO	International Organization for Standardization
ITU-T	International Telecommunication Union-Telecommunication Standardization Sector
VDE	Verband Deutscher Elektrotechniker e.V.

15.2 Type Codes for Communications Cables with Optical Waveguides

A-	Outdoor cable
ASLH-	Self-supporting aerial telecommunication cable for high voltage lines
AT-	Outdoor fan-out cable
B	Multifiber loose buffer
... B ...	Attenuation coefficient and bandwidth at a wavelength of 850 nm
B	Armoring
(1B ...)	One layer of steel tape; ... thickness of steel tape in mm
(2B ...)	Two layers of steel tape; ... thickness of steel tape in mm
Bd	Unit stranding
Cu	Copper wire
D	Multifiber loose buffer, filled
(D) 2Y	Laminated sheath of polyethylene and plastic barrier foil
E	Single-mode fiber
F	Flat constructional design for installation wires
F	Continuous cable core filling
... F ...	Attenuation coefficient and bandwidth at a wavelength of 1300 nm
(F ...)	Flat wire, ... thickness of the wire in mm
FR	Cable with improved flame retardance
G	Graded index fiber
H	Single-fiber loose buffer
... H ...	Attenuation coefficient and bandwidth at a wavelength of 1550 nm
H	Insulating cover or sheath of halogen-free material
J-	Indoor cable
(L) (ZN) 2Y	Laminated sheath made of polyethylene, Al-tape and with nonmetallic strength member
(L) Y	Laminated sheath with polyvinylchloride (PVC)
(L) 2Y	Laminated sheath with polyethylene
Lg	Stranding in layers
NC	Noncorrosive combustion gases
oe	Oil proof
P	Twisted in pairs
P	Step index fiber plastic/plastic
(R ...)	Round wire; ... diameter of the wire in mm
S	Metallic element in the cable core
S	Step index fiber glass/glass
T	Supporting element of steel, textile or plastic
(T)	Nonmetallic concentric supporting element

(TR ...)	Supporting element of round wire; ... diameter of the wire in mm
V	Tight buffered fiber
vzk	Galvanized
W	Single-fiber loose buffer, filled
Y	Insulating cover, sheath or protective cover of polyvinyl chloride (PVC)
2Y	Insulating cover, sheath or protective cover of polyethylene (PE)
02Y	Insulating cover of cellular polyethylene (PE)
02YS	Insulating cover of cellular polyethylene (PE) with additional skin of solid polyolefine (foam skin)
4Y	Insulating cover, sheath or protective cover of polyamide (PA)
6Y	Insulating cover, sheath or protective cover of perfluorethylene propylene (FEP)
9Y	Insulating cover, sheath or protective cover of polypropylene (PP)
11Y	Sheath or protective cover of polyurethane (PUR)
(Z)	Tensile-proof braiding of steel or polyamide wires
(Zg)	Strain relief of glass yarns
(Zn)	Sheath strain relief with nonmetallic elements
(ZN) 2Y	Polyethylene sheath with nonmetallic strength member

15.3 Greek Alphabet

Α α	Alpha	Ι ι	Jota	Ρ ϱ	Rho
Β β	Beta	Κ ϰ	Kappa	Σ σ ς	Sigma
Γ γ	Gamma	Λ λ	Lambda	Τ τ	Tau
Δ δ	Delta	Μ μ	My	Υ υ	Ypsilon
Ε ε	Epsilon	Ν ν	Ny	Φ φ	Phi
Ζ ζ	Zeta	Ξ ξ	Xi	Χ χ	Chi
Η η	Eta	Ο ο	Omikron	Ψ ψ	Psi
Θ ϑ	Theta	Π π	Pi	Ω ω	Omega

15.4 Symbols Frequently Used in Equations

A	Amplitude; cross-sectional area
a	Displacement of a plane wave; core radius; Weibull constant; parameter for calculation for stranding elements
a_K	Attenuation of a cable plant
a_R	Attenuation of repeater spacing
a_{sp}	Splice loss
B; B_1	Bandwidth; bandwidth of an optical fiber with the length L_1
b	Weibull constant; parameter for calculation of stranding elements
b_1	Bandwidth length product
c	Concentricity (core/cladding); parameter for calculation of stranding elements; speed of light in a medium
c_g	Group velocity
c_0	Speed of light in a vacuum
D; D_{min}; D_{max}	Diameter; cladding diameter (min. and max.)
d; d_{min}; d_{max}	Diameter; core diameter (min. and max.)
D_0	Cladding diameter if circular
D_1	Total diameter of the stranding elements
d_0	Core diameter if circular
d_1; d_2	Diameter of stranding elements
d_i	Inside diameter of a loose buffer
d_f	Outside diameter of the optical fiber

E	Young's modulus; cladding nonconcentricity
E_i	Young's modulus of the material i
E_g	Energy gap
e	Core noncircularity; Euler's number = 2.7182818 ...
F	Tensile force; probability of fracture
F_{max}	Maximum tensile force
f	Frecuency
f_m	Frequency of modulation
G	Weight; torsion modulus
g	Profile exponent
$g_1(t)$; $g_2(t)$	Input/output pulse as a function of the time t
$H(f_m)$	Transfer function of the frequency of modulation f_m
$H(0)$	Transfer function at the frequency of modulation $f_m = 0$ Hz
$I_N(r)$	Near field light intensity as a function of the radius r
$I_F(\theta)$	Far field distribution as a function of the angle θ relative to the optical axis
$J_0(x)$	Bessel function of 0th order
$J_n(x)$	Bessel function of nth order
k; $\vec{k}$	Wave number; wave vector
L	Length, length of the stranding element
L_0	Length under testing
L_1	Length of an optical fiber with the bandwidth B_1
$M(\lambda)$	Chromatic dispersion
M_0	Material dispersion
$M_0(\lambda)$	Material dispersion at the wavelength λ
M_1	Waveguide dispersion
$M(\lambda_m)$	Dispersion at the medium wavelength λ_m of the transmitter
$M'(\lambda_0)$	Slope of the dispersion at the wavelength λ_o
N	Number of modes guided in the core
NA	Numerical aperture of the optical fiber
n	Refractive index = phase refractive index; number of stranding elements; number of splices
n_g	Group refractive index
n_0	Refractive index = phase refractive index in a vacuum ($\approx$ air)

P	Profile dispersion
$P(L)$, $P(L_1)$, $P(L_2)$	Light power in the optical fiber after the length L, L_1, or L_2
$P(0)$	Light power at the near end of the optical fiber
$P_1(f_m)$, $P_2(f_m)$	Amplitude of the light power at the near/far end of the optical fiber as a function of the frequency of modulation f_m
R	Stranding radius
ΔR	Free space of the optical fiber in a buffer tube
r	Radius; distance from the axis of the optical fiber
S	Lay length
T	Period of oscillation; pulse duration; temperature
ΔT	Temperature change
ΔT_{FRMS}	Full root mean square pulse broadening
t	Time
t_g	Group transit time
t_0	Nominal duration of test
V	V number or normalized frequency
V_c	Limit value of the V number
w_0	Mode field radius
x	Deviation from core center
Z	Excess length due to stranding
z	Length in the z direction
z_0	Fixed point of a wave in z direction
α	Angle of incidence of the light; linear thermal expansion coefficient; stranding angle; scattering loss; attenuation coefficient (attenuation constant)
α_0	Angle of incidence with the angle of refraction β_0
α_{res}	Extra margin of attenuation
β	Angle of refraction
β_0	Angle of refraction at the angle of incidence α_0
γ	Density of the material; gamma factor
Δ	Normalized refractive index difference
$\Delta\lambda$	Spectral full width half maximum (FWHM)
$\Delta\lambda_{FRMS}$	Spectral full root mean square width (FRMS)

ε	Elongation; ellipticity
ε_K	Cable elongation
ε_{TK}	Cable contraction
ε_F	Optical fiber elongation
η	Viscosity
Θ	Angle relative to the axis of the optical fiber
Θ_{max}	Acceptance angle of the optical fiber
λ	Wavelength
λ_c	Cut-off wavelength
λ_m	Medium wavelength of a transmitter
λ_0	Wavelength at which the chromatic dispersion is zero
μ	Radial mode number; Poisson's ratio
ν	Azimuthal mode number
π	Pi = 3.1415927 ...
ϱ	Bending radius
σ	Stress; breaking stress
σ_0	Nominal stress
φ	Phase angle
ω	Angular frequency

15.5 Conversion Tables

Length

Unit	Meter	Inch	Foot	Yard
1 m	1	39.3701	3.2808	1.0936
1 in	0.0254	1	0.0833	0.0278
1 ft	0.3048	12	1	0.3333
1 yd	0.9144	36	3	1

Unit	Kilometer	Mile
1 km	1	0.6214
1 mile	1.6093	1

Area

Unit	Square meter	Square inch	Square foot
1 m^2	1	1550.0	10.7639
1 in^2	$0.6452 \cdot 10^{-3}$	1	$6.9444 \cdot 10^{-3}$
1 ft^2	0.0929	144	1

Volume

Unit	Cubic decimeter or liter	Cubic inch	US gallon
1 $dm^3 = 1\ l$	1	61.0237	0.2642
1 in^3	0.0164	1	$4.3290 \cdot 10^{-3}$
1 US gal	3.7854	230.9991	1

Velocity

Unit	Meter per second	Kilometer per hour	Miles per hour	Foot per second
1 m/s	1	3.6	2.2369	3.2808
1 km/h	0.2778	1	0.6214	0.9113
1 mile/h	0.4470	1.6093	1	1.4667
1 ft/s	0.3048	1.0973	0.6818	1

Mass

Unit	Kilogram	Pound	Ounce
1 kg	1	2.2046	35.274
1 lb	0.4536	1	16
1 oz	$28.3495 \cdot 10^{-3}$	0.0625	1

Mass Relative to length

Unit	Kilogram per kilometer	Tex	Denier	Pound per 1000 feet
1 kg/km	1	10^3	9×10^3	0.671969
1 tex	10^{-3}	1	9	$0.6720 \cdot 10^{-3}$
1 den	$0.1111 \cdot 10^{-3}$	0.1111	1	$74.6631 \cdot 10^{-6}$
1 lb/kft	1.488164	1488.164	13393.48	1

Pressure

Unit	Newton per square millimeter	Pound force per square inch (psi)
1 N/mm^2 = 1 Mpa	1	145.038
1 lbf/in^2 = 1 psi	6.8948×10^{-3}	1
Other unit: 1 Pa = 1 N/m^2 = 10^{-6} N/mm^2 = 10^{-5} bar		

Force

Unit	Newton	Kilopond	Pound force
1 N	1	0.101972	0.22481
1 kp	9.80665	1	2.20462
1 lbf	4.44822	0.45362	1

Power

Unit	Watt	Kilocalorie per hour	British thermal unit per hour
1 W	1	0.8598	3.4121
1 kcal/h	1.163	1	3.9683
1 Btu/h	0.2931	0.2520	1

Work, Energy

Unit	Kilojoule	Kilocalorie	British thermal unit
1 kJ	1	0.2388	0.9478
1 kcal	4.1868	1	3.9683
1 Btu	1.0551	0.2520	1

Temperature

Conversion of degrees Celcius T_C into degrees Fahrenheit T_F:

$$T_F = \frac{9}{5} \cdot T_C + 32$$

Conversion of degrees Fahrenheit T_F into degrees Celsius T_C:

$$T_C = \frac{5}{9} \cdot (T_F - 32)$$

16 Glossary

A

absorption
Loss of radiation during transit through a medium, whereby, for example in the case of → light part of the radiation energy is transformed into heat and other forms of energy. Because the molecules of the material through which the radiation passes can only take up specific amounts of energy in the form of excitation energy, only certain → wavelengths are absorbed, depending on the properties of the material. In the case of OH ions in an optical fiber, this is the wavelength of 1390 nm. In → photodiodes, → absorption is the process by which the energy of an incident photon raises an electron from the valence band to the conduction band.

acceptance angle
Largest possible → launch angle θ_{max}, also known as angle of aperture, within which → light can be launched into the → core of an optical fiber and guided in it. If this angle is exceeded during launch, then → leaky modes are created or → modes that cannot propagate in the fiber.

alignment
Optimal positioning of the ends of the → optical fiber for → splice joints. In the case of fusion splicing of single-mode fibers, alignment of the optical fibers can be carried out with a → LID system.

amorphous
Structure of a solid body which, in contrast to crystals, is composed of irregularly arranged atoms or molecules, e.g. → fused silica glass, amber.

annealing point
Upper temperature boundary of the transformation range. For → fused silica glass the upper annealing point is 1180 °C, whereby the viscosity $\eta = 10^{13}$ dPa · s or $\log \eta = 13$).

armoring
Protective element (usually of steel wires or steel bands) used in cables for special applications, e.g. for submarine or mine applications, for cables with rodent protection, etc. It is applied over the → cable sheath.

attenuation
Reduction in the optical power between two cross-sections of an → optical fiber. It is dependent on the → wavelength. Its main causes are → scattering and → absorption as well as light losses due to → connectors and → splices. It is described with the logarithmic measure

$$-10 \log \frac{P(L_1)}{P(L_2)} \text{ in dB.}$$

$P(L_1)$ light power in the cross-sectional area at the length L_1

$P(L_2)$ light power in the cross-sectional area at the length L_2

attenuation coefficient
Attenuation with respect to length of a uniform → optical fiber in → steady state (unit: dB/km).

avalanche photodiode
Receiver element based on the avalanche effect, i.e. the photocurrent is amplified with low noise by carrier multiplication in an electric field.

azimuthal mode number
Number for identification of the propagative → modes in an → optical fiber. The azimuthal mode number gives half the number of light points in each concentric ring (Figure 2.9).

B

backscattering technique
Technique for measurement of the → attenuation along an → optical fiber. Most of the light power propagates forward, but a small percentage is scattered backward toward the → transmitter. By observing the time curve of the backscattered → light with a beamsplitter at the transmitter, it is possible to measure from one end not only the length and → attenuation of the installed homogeneous optical fiber but also local irregularities, fractures and optical power losses in → splices and → connectors.

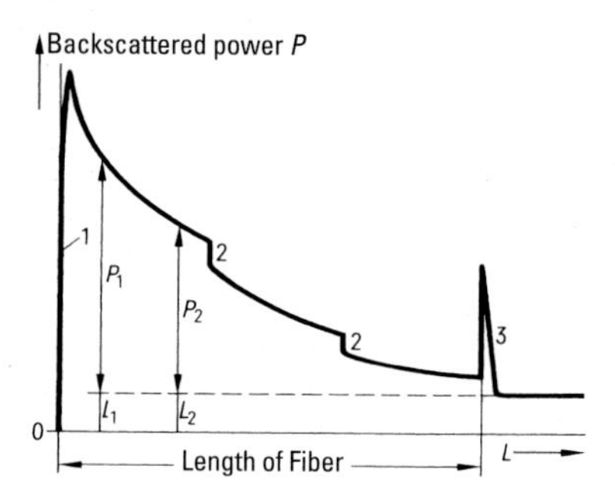

band gap
Energy gap between valence band and conduction band of a → semiconductor. This energy interval cannot be occupied by electrons. In optoelectronic semiconductor components, the energy gap determines their operating → wavelength.

bandwidth
Frequency at which the magnitude of the → transfer function of an optical fiber has fallen to half its value at zero frequency, i.e. the frequency at which the signal attenuation has increased by 3 db. Since the bandwidth of an optical fiber is approximately reciprocal to its length (→ mode mixing), the → bandwidth length product is often given as a characteristic value.

bandwidth length product
With negligible → mode mixing (→ gamma factor = 1) the → bandwidth of an → optical fiber is inversely proportional to its length and therefore the product of bandwidth and length is constant.

bending method
Measurement technique for determining the → cut-off wavelength of a → single-mode fiber. Hereby the additional → attenuation caused in a 2 m long piece of the → optical fiber by winding it around a mandrel, e.g. with a diameter of 30 mm, is measured. The cut-off wavelength is then the → wavelength at which the maximum of the additional → attenuation with the longest wavelength equals the value 0.1 dB.

Bessel function
Mathematical function for the description of the electric field in cylindrical waveguides, such as coaxial cables, hollow tubes or → optical fibers. The

Bessel functions $J_n(x)$ look similar to dampened sine oscillations (Figure 4.2). For fibers with → step index profile, the zeros of these functions indicate the value of the → V number at which the respective → modes have their → cut-off wavelength.

B-ISDN
→ Integrated Services Digital Network (ISDN) for transmission of a large variety of broadband services (high → bit rates), such as video phone, work place and studio video conferences, broadband videotext, fast data transmission, interactive cable television.

bit
Basic unit of information in digital transmission systems. A bit is the numeral for counting binary signals; it is equivalent to the decision between two states, usually described as 0 or 1. In digital electronics bits are represented by pulses. A group of eight bits is usually called a byte.

bit error rate (BER)
Ratio of the average number of bit errors occurring over a longer period of time during digital signal transmission to the number of bits → transmitted in this period of time.

bit rate
Transmission speed of a binary signal, with → bits following each other in a rigid time frame.

boron trioxide (B_2O_3)
Chemical compound used for → doping in production of → optical fibers in order to lower the → refractive index.

branching element
Fiber optic component for dividing the → light power between several incoming → optical fibers and several outgoing fibers (→ coupler); also known as tapping element.

bus network
Network in which the individual stations (subscribers) are connected via a transceiver to the bus line with corresponding terminating resistors.

butt joint
Coupling of two → optical fibers or diodes whereby their exit and entrance surfaces lie parallel and in close proximity to one another.

C

cable core
In a cable, all of the stranding elements, antibuckling and strength members, fillers and the wrapping enclosing them.

cable sheath
Sheath or jacket usually of polyethylene (PE) or polyvinylchloride (PVC) which protects the → cable core from environmental influences.

chromatic dispersion
Combination of the two related effects → material dispersion and → waveguide dispersion.

circularly polarized light
→ Light for which the pointer of its vector of the electric or magnetic field describes a circle.

cladding
All of the optically transparent material in an → optical fiber, with the exception of the → core.

cladding diameter
Diameter of the smallest circle that encompasses the cross-section of the → cladding. Cladding radius is the radius of this circle.

cladding glass
Material of the → cladding, made of glass with a lower → refractive index relative to the → core glass.

coating
The plastic protective layer applied directly to the surface of the → cladding during production of the → optical fiber.

coding
Transformation of information signals for transmission as signal elements in steps (→ modulation rate). Different values and conditions can be represented by each signal element. For example, a binary signal has two conditions (→ bit), a ternary signal three conditions, a quaternary signal four conditions, etc. Examples of typical coding in binary signals can be seen in the table at the bottom of this page.
NRZ is the abbreviation for nonreturn-to-zero and RZ stands for return-to-zero. A further example of coding are the general nBmB codes with a block of information of *n* bits which is trans-

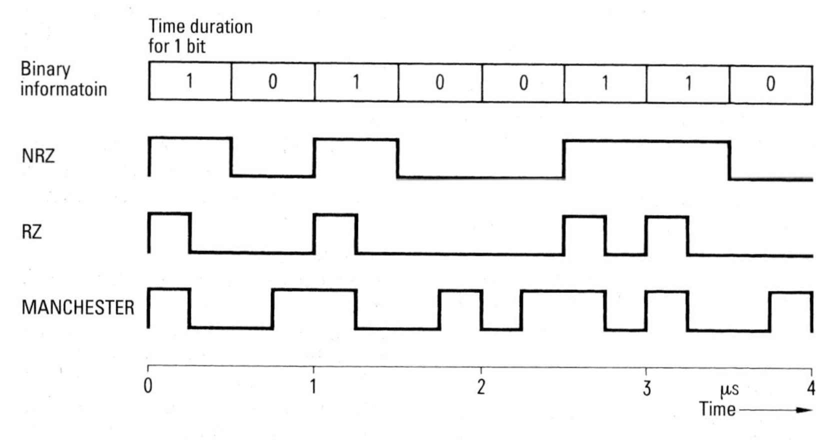

Code	Description	Bit rate (Mbit/s)	Modulation rate (MBaud)	Duration of step (µs)
NRZ	Binary 1 ≙ high Binary 0 ≙ low	2	2	0.5
RZ	Binary 1 ≙ high for half of the bit period Binary 0 ≙ low	2	4	0.25
MANCHESTER	Binary 1 ≙ transition from high to low at half the bit period Binary 0 ≙ transition from high to low at half the bit period	2	4	0.25

formed into m bits for transmission where m is larger than n.

coherence length
Propagation distance along which a → light wave can be described as coherent, i.e. it remains unchanged in phase and → wavelength.

coherent light source
Optical source which emits → coherent waves. If the spectral width of the optical source is $\Delta\lambda$ and the medium wavelength λ_m, then the → coherence length of the waves transmitted in a medium with the → refractive index n is approximately $\frac{\lambda_m^2}{n \cdot \Delta\lambda}$.

coherent waves
Waves with identical → wavelengths and constant phase difference in time relative to each other (→ interference).

composite buffered fiber
Combination of → single-fiber loose buffer and → tight buffered fiber. The free space between the → optical fiber and the buffer is reduced to a minute space and filled with a gliding layer.

connector
Easily disconnected joint of two → optical fibers. Usually the → insertion loss is higher with connectors than with splice joints.

consolidating
Shrinking of the porous soot layers during production of a preform to form a solid transparent glass rod, which can then be drawn out to form a → fiber.

core
Central part of an → optical fiber, which serves as a waveguide.

core diameter
Diameter of the smallest circle which encompasses the area of the → core cross-section. The core radius is the radius of this circle.

core glass
Material of the → core with higher → refractive index relative to the → cladding glass.

coupler
Passive optical component for transmission of → light between optical source and → optical fiber or between several optical fibers. Couplers are particularly important because they make it possible to connect several → transmitters and → receivers (→ star coupler, → branching element, tapping element) in optical networks.

crimp
Creation of a good mechanical connection by permanently deforming a sleeve around an optical fiber buffer.

critical angle
The angle of incidence of a light ray at the transition from a material with a higher → refractive index n_1 to a material with a lower refractive index n_2 at which the angle of refraction is 90° and the light is no longer guided. The sine of the critical angle a_0 is equal to the ratio of the two refractive indices:

$$\sin \alpha_o = \frac{n_2}{n_1}.$$

curve of the refractive index
Graphic representation of the → refractive index as a function of the → wavelength.

cut-back method
One of the possible methods to measure → attenuation, whereby the opti-

cal power is measured at two points (L_1 and L_2) of the → optical fiber. Point L_2 is at the far end of the optical fiber and L_1 is very close to the near end of the optical fiber, i.e. this method is problematical because it requires that a short piece of the optical fiber be cut. By contrast, → the insertion loss technique is not destructive.

cut-off wavelength
Shortest → wavelength at which only the → fundamental mode of an → optical fiber is capable of propagation. For → single-mode fibers the cut-off wavelength must be smaller than the wavelength of the light to be transmitted.

destructive interference
→ Interference of two → coherent waves with the same amplitude and with a phase difference of an uneven multiple of half of the → wavelength.

DFB (distributed-feedback) laser
→ Laser diode with a spectral width of less than 1 nm which has a series of wavelike elevations on the semiconductor substrate to reflect only certain → wavelengths, thereby amplifying a single resonance wavelength.

dielectric medium
Insulating material, i.e. one which does not conduct electricity.

differential mode delay measurement
Determination of delay time of individual groups of → modes in a → multimode fiber.

dispersion
Spreading of the group delay time in an → optical fiber. It is a sum of a variety of different parts: → modal dispersion, → material dispersion, and → waveguide dispersion. Due to the dispersion, → light pulses in an optical waveguide broaden temporally. As a result of this effect, the optical waveguide acts as a low-pass filter for the transmitted signals (→ transfer function).

distillation
Method for separating chemical materials, by which they are heated to boiling in a container and then, by cooling their vapors in a tube with a downward slope, they are collected again, the condensed liquid (distillate) being collected in a second container.

distributed-feedback laser
→ DFB laser

dopant concentration
Material concentration of the → doping relative to the base material, usually given in ppm (parts per million) or ppb (parts per billion), i.e. one foreign particle per million or billion particles of the base material.

doping
Defined addition of small amounts of a foreign substance to a pure material, in order to slightly change its properties. For example the increased → refractive index of the → core of an → optical fiber is created by doping the base material, → silicon dioxide, with → germanium dioxide.

double crucible method
→ Core and → cladding glass are melted in two separate crucibles, united in the molten state and immediately afterwards they are drawn out to a → fiber.

double heterostructure
Sequence of layers in an optoelectronic

→ semiconductor component, in which the active semiconductor layer is sandwiched between two covering layers with a higher energy gap. In a → laser diode the double heterostructure has a dual function: confinement of the charge carriers and dielectric waveguiding.

E

eigenwelle
→ modes.

electric interface
Interface or transition from one electrical circuit to another. The definition of the electrical parameters, e.g. voltage and resistance at an electrical interface, is critical for the proper interaction of the two bordering circuits.

electromagnetic wave
Periodic changes in the state of the electromagnetic field that propagate in waves at the speed of → light. In the range of the optical frequencies they are called light waves.

elliptically polarized light
→ Light, for which the tip of the vector of its electrical or magnetic field describes an ellipse.

erbium Er
Chemical element, number 68 in the periodic system, which belongs to the lanthanide family. It is used, among other things, to → dope single-mode optical fibers for use in → optical fiber amplifiers.

excess length due to stranding
Extra length of the → stranding elements as compared to the length of the cable, due to the fact that they do not lie parallel to the axis of the cable.

extruder
Machine for manufacture of seamless thermoplastic jackets (e.g. → cable sheath, buffer, etc).

F

far field distribution
Distribution of the radiated optical power from a → light source or from the end of an → optical fiber over an angle relative to the optical axis. It is critical for the mode field distribution at greater distances from the light source and in optical fibers it is dependent on the length of the fiber and the → launch conditions, as well as on the → wavelength.

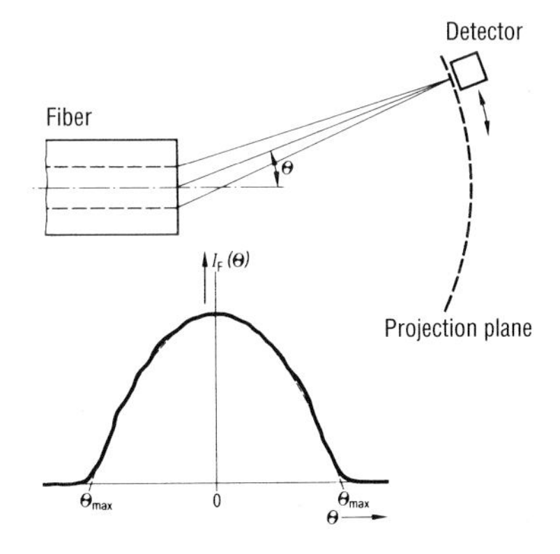

FDDI
fiber distributed data interface
High speed network (100 Mbit/s) utilizing → optical fibers.

fiber
→ optical waveguide.

fiber multiplex
Transmission method, in which one → optical fiber is assigned to each transmission channel. In contrast to this, there are → time division multiplex and → wavelength division multiplex (WDM).

FITL
Fiber-in-the-Loop
→ Optical fibers in the local subscriber access network, that part of the public network which connects subscribers to an exchange. Depending on the end point of the optical fiber route FTTH (fiber-to-the-home), FTTC (fiber-to-the-curb) and FTTP (fiber-to-the-pedestal) are differentiated.

fluorine (F)
Nonmetallic, gaseous chemical element that is used as a → dopant in the manufacture of → optical fibers to reduce the → refractive index.

four-circle method
Method in which a mask is used to check the geometric tolerances of an → optical fiber. It is not suitable for determination of exact values of the diameter, circularity and concentricity.

frequency domain method
Method for measurement of the → bandwidth of an → optical fiber in the frequency range.

frequency of modulation
Frequency of the → modulation f_m, by which the → amplitude of the optical power of the light from a → transmitter which is to be launched into an → optical fiber is varied (modulated).

full flood launch
→ Launch condition in which both the → numerical aperture and the light spot (homogeneous power density) of the → transmitter is equal to or greater than the corresponding → core parameters of the → optical fiber, which leads to excitation of all → modes, including → leaky modes.

fundamental mode
Lowest order → mode in an → optical fiber. It usually has a bell-shaped (Gaussian) mode field distribution and its dimension is described in the → near field by the → mode field diameter. In → single-mode fibers operated above the → cut-off wavelength it is the only propagative mode.

fused silica glass
An → amorphous, i.e. noncrystalline, glassy solidified melt of → silicon dioxide (SiO_2) which, due to its high → viscosity, appears to be solid. The crystalline form of SiO_2 is called quartz.

G

gamma factor
Because of → modal dispersion, the → bandwidth decreases nonlinearly when several lengths of cable are concatenated; therefore the gamma factor is used to calculate an approximation of the actual bandwidth length dependence in → multimode fibers.

Gaussian low pass filter
Filter (e.g. for → frequencies of modulation), in which the output amplitude (of → the modulation) exhibits a Gaussian-shaped decrease (roll-off) as the frequency increases. The typical → transfer function of an → optical fiber has a curve comparable to a Gaussian low-pass filter.

germanium dioxide (GeO_2)

Chemical compound which is used most frequently in the manufacture of → optical fibers as a → dopant, in order to increase the → refractive index.

glass doping

→ doping.

graded index profile

→ Refractive index profile of an → optical fiber which decreases continuously from the inside toward the outside (usually parabolically) (→ profile exponent $1 \leq g \leq 3$).

group index

Factor n_g by which the propagation speed of a finite wave group (e.g. a → light pulse), i.e. the → group velocity, in an optically more dense medium is smaller than it is in a vacuum. The group index n_g for a → wavelength λ can be calculated from the → phase refractive index n:

$$n_g = n - \lambda \frac{dn}{d\lambda}.$$

It is the group index which should be used to describe the propagation of a light pulse in an → optical fiber (→ refractive index).

group velocity

Propagation speed of a wave group (e.g. of a → light pulse), which consists of individual waves with different → wavelengths (→ group index).

GRP element

Antibuckling and strength member made of glass filaments (GRP glass fiber reinforced plastic).

H

helical stranding

→ Stranding method, by which the → stranding elements of one layer are stranded in one direction with a constant angle relative to the longitudinal axis of the cable.

Hooke's law

In solid bodies small deformations are proportional to the external forces and therefore to the internal tensions, e.g. the tensile stress σ is proportional to the elongation ε (lengthening or shortening per unit of length), with → Young's modulus as the proportionality constant $\sigma = E\varepsilon$.

I

ICCS integrated communications cabling system

Structured, service independent cabling system developed by Siemens for the subscriber and backbone areas in which only two different cable types are used. → Optical fiber cables for the backbone and campus area (primary area), as well as for the riser areas in buldings (secondary area), and mainly balanced copper cables for the subscriber area, the individual floors (tertiary area).

indoor cable

Cable for a wide variety of applications inside a building. It is not designed for outdoor installation.

infrared radiation

Range of the spectrum of → electromagnetic waves from abut 780 nm (red) to the longer → wavelengths of about 1 mm. The wavelengths preferred today for → optimal communication technology lie in the near-infrared range, whereby 850, 1300 and 1550 nm are the preferred wavelengths.

insertion loss
Attenuation caused by the insertion of an optical component into an optical transmission route (e.g. → connectors or → couplers).

insertion loss technique
One of the possible methods for measurement of the → attenuation, in which the optical power at the end of the → optical fiber under test is measured, in order to compare it with the optical power at the end of a short piece of fiber. This piece of fiber serves as a reference and should correspond to the optical fiber under test.

inside vapor deposition (IVD) method
Method for manufacture of → optical fibers in which glass is deposited out of the vapor phase onto the inside of a rotating glass tube.

integrated services digital network (ISDN)
Unified digital network, in which all types of communication (voice, text, data, stationary and moving picture) are transmitted from one port (terminal) in the exchange (switch) over one access line to and from the subscriber.

intensity
Power concentration on surface through which radiation passes, e.g. on the radiating surface of an optical source or on the cross-sectional area of an → optical fiber (usual unit: W/cm^2); also called radiant flux or flow irradiance.

interference
Superposition of two or more waves passing the same location at the same time. The resulting amplitude is equal to the sum of the amplitude of the original waves respectively. Only in the case of → coherent waves is this effect particularly marked.

L

LAN (Local Area Network)
Local network for bit serial transmission of information between interconnected independent terminals. The term LAN was introduced to differentiate between processor coupling (central unit with peripheral units) with very high → bit rates over short distances, on the one hand, and the public networks over longer distances, on the other hand.

laser diode
Transmitter diode which emits coherent → light above a threshold current (→ stimulated emission). A differentiation is made between gain guided and index guided laser diodes.

laser module
The operation of → laser diodes requires several different functional circuit boards. They are housed in a hermetically sealed box, called a laser module.

launch angle
Angle between the direction of propagation of the incident → light ray and the axis of the → optical fiber. In order that the incident light can be coupled in, this angle must lie between zero and a maximum value which depends on the position on the end face of the

optical fiber or its local → refractive index difference relative to the → cladding (→ numerical aperture *NA*).

launch conditions
Conditions under which → light is coupled into the → optical fiber. They are important for the further distribution of the light power in the optical fiber.

layer cable
Cable in which the → stranding elements are individual elements which are arranged concentrically in one or more layers, usually around a central member.

lay length
Axial length along the center axis of a cable, after which a → stranding element has completed one full turn (360°) around this axis.

lead alkali silicate glass
Multicomponent glass which is used as a basic raw material for manufacture of → optical fibers. Due to its many impurities, it exhibits high → attenuation.

leaky modes
→ Modes which propagate in the boundary region between the guided → modes of an → optical fiber and the → light waves which are not capable of propagation. Strictly speaking, they are not guided, but are capable of limited propagation with increased → attenuation.

LID (Local Injection and Detection)
The local → light injection and detection system is an aid for quick and unproblematical alignment of → single-mode and → multimode fibers. The entering and exiting → optical fibers in the splicer are wound around a mandrel of sufficiently small diameter that, via a fat core fiber coupled to it at these points, light can be coupled into (light injection) and out of (light detection) the fiber (through the colored coating and the cladding glass). By means of the light transmission from the incoming to the outgoing fiber, aided by a microprocessor, an optimum alignment of the fibers can be achieved for splicing.

light
The term "light" was originally applied only to light visible to human eyes, i.e. → electromagnetic waves with a → wavelength between 380 (violet) and 780 nm (red). However, it has become customary to use the term for radiation in the ranges bordering visible light (e.g. near- → infrared radiation at 800 to about 1600 nm) in order to emphasize their physical and technical similarities to visible light. Therefore, for example, both in English and in German, the terms used for radiating sources like → light emitting diode and → laser contain the word "light". The words related to "light" which stem from the Greek word "optics" and "photo" occur in many word combinations.

light emitting diode (LED)
Semiconductor component which radiates incoherent light due to → spontaneous emission.

limited phase space launch
→ Launch condition in which about 70 per cent of the → core diameter and 70 per cent of the → numerical aperture of an → optical fiber is flooded with the light to be launched into it, with the aid of a combination of lenses and apertures.

linearly polarized light
Light for which the tip of the vector of its electrical of magnetic field describes a straight line.

linear thermal expansion coefficient
The expansion which occurs in a solid body (in one direction), e.g. a cable in the direction of its length, is called linear expansion. A result of temperature, it occurs as a first approximation in proportion to the temperature increase. The linear thermal expansion coefficient is the fraction by which the length increases per degree Kelvin.

M

macrobending
Macroscopic axial deviations of an → optical fiber from a straight line (e.g. on a delivery spool), in contrast to → microbending.

MAN
metropolitan area network
City wide, fast (34, 45, 140 Mbit/s) communications system, which enables data transmission in addition to voice applications.

material dispersion
→ Dispersion due to the → wavelength dependence of the → refractive index n of a material. It is characterized by the material dispersion M_0 (λ). For the majority of → optical waveguide materials M_0 becomes zero at a certain wavelength λ_0 in the vicinity of 1300 nm.
The usual unit for M_0 is $\frac{\text{ps}}{\text{nm} \cdot \text{km}}$.

Maxwell's equations
Equations developed by C. Maxwell to describe all electromagnetic phenomena, to which → light also belongs.

microbending
Bending of the → optical fiber with spatial → wavelengths of a few millimeters, for example, and local axial deviations of a few micrometers. It causes light losses and thereby increases the → attenuation of the optical fiber.

modal dispersion
→ Dispersion in an → optical waveguide caused by the superpositioning of → modes with different delay times at the same wavelength. It is also called multimode distortion.

modal noise
Amount of noise in optical transmission systems, which is created by a combination of mode-dependent → attenuation and varying distribution of the spatial energy and the relative phases of the guided → modes. It is more severe when the coherence of the radiation of the → laser diode is high. Improvcmcnts in the technology methods of connectors and splice joints and application of → light emitting diodes or wide spectrum multimode laser diodes and most importantly introduction of → single-mode fibers have alleviated it. It is also called speckle noise.

mode field diameter
In order to characterize the → light distribution of the → fundamental mode in → single-mode fibers the mode field diameter $2w_0$ or the mode field radius w_0 is used.

mode filter
If it is desirable to suppress higher → modes, when → light is launched into a → multimode fiber, then a mode filter is used; e.g. the optical fiber in question is wound around a mandrel with a diameter of about one centimeter, which causes higher modes to be stripped. In general, it can be stated that mode filters are used to limit exci-

tation to certain modes (mainly lower order modes).

mode mixing
The many → modes of a → multimode fiber differ among other parameters in their velocity of propagation. Due to perturbations of the → optical fiber geometry and the → refractive index profile, a gradual energy exchange occurs between the modes of different speeds. As a result of this → mode mixing the → bandwidth of long multimode fiber is higher than that linearly extrapolated from measurement of a short optical fiber.

mode scrambler
Device with which the → steady state can be approximated in a → multimode optical fiber by causing strong → mode mixing in a short piece of optical fiber through statistically irregular mechanical perturbations. For this purpose, for example, a short piece of fiber can be pressed against a rough surface (sand paper, etc).

mode stripper
Device for stripping → cladding → modes. In → single-mode fiber, for example, the → coating has the effect of a → mode stripper. During → full flood launch → leaky modes and cladding modes are excited, which, however, are stripped within a few centimeters when a coating is used with a → refractive index higher than that of the → cladding glass.

mode volume
The product of the cross-sectional area and the solid angle in an → optical fiber that is available for propagation of → light in an optical fiber. It is proportional to the number of propagative → modes (→ phase space diagram). In → multimode fibers, the guided light power is usually not distributed evenly over all propagative modes, e.g. because these did not have the same → launch conditions or because the modes are attenuated differently. Thereby, the use of the mode volume as well as the extent of the → near field distribution and → far field distribution of the optical fiber are reduced.

modes
Solutions of → Maxwell's equations, taking into consideration the border values related to waveguides, i.e. an → eigenwelle, which has a transversal field distribution independent of the direction of propagation.

modified chemical vapor deposition (MCVD) method
→ Inside vapor deposition method, whereby the energy needed for glass deposition is provided by an oxihydrogen gas burner.

modulation
Change in a parameter of a wave (e.g. amplitude, frequency, phase) in order to transmit information by this wave. In amplitude modulation (AM) the amplitude of the carrier oscillation is influenced by the oscillation of the information. In frequency modulation (FM) the frequency of the carrier is changed by the timing of the communication. In → pulse code modulation (PCM) the carrier is radiated in the form of short oscillations (pulses) which contain the coded information.

modulation rate
→ Frequency of modulation of a digital transmission system, which is equal to the reciprocal of the duration of the shortest pulse, measured in seconds, i.e. pulse width of a communications signal. It is also called the signal speed. Its unit is the baud:

$$1 \text{ baud} = \frac{1}{\text{pulse width in s}}.$$

The modulation rate is not dependent on the format of the data, i.e. the method of → coding the data.

monomode optical waveguide
→ single-mode fiber.

multifiber loose buffer
It consists of several → optical fibers in a common loose buffer (→ single-fiber loose buffer).

multimode fiber
→ Optical fiber with a large → core diameter compared to the → wavelength of the → light guided and in which a large number of → modes are therefore propagative. By means of a → graded index profile the → modal dispersion can be kept to a minimum, so that higher → bandwidths can be archieved, which, however, are still exceeded by those in → single-mode fibers.

N

near field distribution
Distribution of the → intensity of the → light over an irradiated surface, e.g. over the cross-sectional area of an → optical fiber.

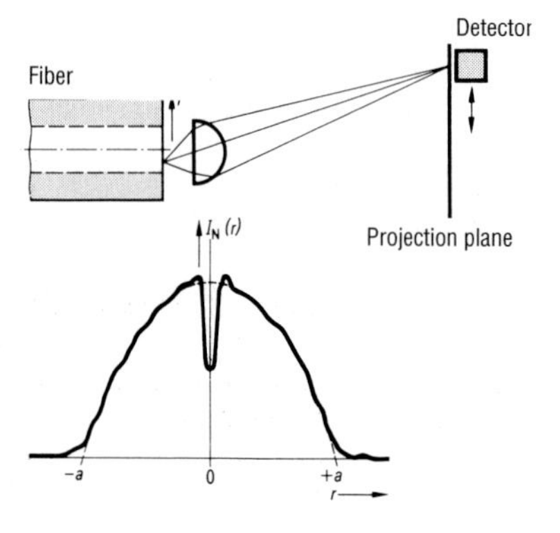

numerical aperture
Sine of the → acceptance angle θ_{max} of an → optical fiber, which only depends on the → refractive indices of the → core n_1 and the → cladding n_2:

$$NA = \sin \theta_{max} = \sqrt{n_1^2 - n_2^2}.$$

O

optical communication
Technology for transmission of communication signals with the aid of → light.

optical fiber
→ optical waveguide.

optical fiber amplifier
→ Erbium-doped single-mode fiber, which uses pump light from a laser diode for direct optical amplification of light signals.

optical time domain reflectometer (OTDR)
Device for the measurement of → attenuation of an → optical fiber which is based on the → backscattering technique.

optical transmission systems
Systems (technical configurations and components) which are used in → optical communications technology to transmit communication signals.

optical waveguide (OWG)
Dielectric waveguide with a → core consisting of optically transparent material of low → attenuation and with a → cladding consisting of optically transparent material of lower → refractive index than that of the core. It is used for transmission of signals by → electromagnetic waves in the range

of optical frequencies. The → optical waveguide usually has a protective → coating. It is often referred to a → optical fiber.

outdoor cable
Cable with structure and dimensions designed, such that it meets all requirements for buried, duct or aerial cable plants. It usually has a PE sheath.

outside vapor deposition (OVD) method
Method for manufacture of → optical fibers by glass deposition out of the vapor phase onto the outside surface of a rotating → target rod.

P

passive optical network
→ PON

phase refractive index
Factor *n* by which the velocity of propagation of an infinite, unmodulated optical source, i.e. the phase velocity in an optically dense medium is smaller than that in a vacuum.

phase separable glass method
Method for manufacture of → optical fibers with an → attenuation coefficient of 10 to 50 dB/km at 850 nm. Basic material for this method is a rod of → sodium borosilicate glass that is heated and leached out. Cesium can be used for → doping.

phase space diagram
Graphic representation of the → light guiding properties of an → optical fiber in a suitable system of coordinates. A variety of optical fiber properties can thus be described, e.g. optical power which can be coupled into a fiber, → launch conditions, etc.

photodiode
Diode made of → semiconductor material which absorbs → light and guides the photo current of charge carriers freed to a external electrical circuit. A differentiation is made between → PIN photodiodes and → avalanche photodiodes.

piezoelectricity
Electrical charging of some crystals by means of mechanical pressure. Piezoelectric crystals are used in high frequency technology as quartz-controlled → transmitters (crystal markers) and as electrical filters. They are also used to measure mechanical deformations, high pressures and as electromechanical transducers (acoustical oscillators, positioning elements, microphones, etc.).

pigtail
Short piece of an → optical fiber used to couple optical components, e.g. a → laser diode with a → connector. It is usually permanently attached to the component.

PIN photodiode
Receiver diode (→ photodiode) with main → absorption in a depletion region (i region) within its pn junction. In contrast to an → avalanche photodiode, such a diode has a high → quantum efficiency but no internal current amplification (avalanche effect).

plasma-activated chemical vapor deposition (PCVD) method
→ Inside vapor deposition method, whereby the energy for the glass deposition is applied by means of a plasma (an ionized gas).

PON
passive optical network
Optical network for → FITL in the local subscriber access network, based on passive optical components, e.g. → couplers, → splitters, → connectors as well as on → fiber optic cables.

Poisson's ratio
Ratio of diameter contraction to elongation of a body under the influence of tensile force. Poisson's ratio can be calculated using → Young's modulus and the → shear modulus.

power law index profile
→ Refractive index profile described as a power function of the radius r:

$$n^2(r) = n_1^2 \cdot \left[1 - 2 \cdot \Delta \left(\frac{r}{a}\right)^g\right]$$

for $r < a$, in the core

$$n^2(r) = n_2^2 = \text{constant}$$

for $r \geq a$, in the cladding

n_1 refractive index at the → optical waveguide axis
n_2 refractive index of the → cladding
Δ normalized → refractive index difference
a → core radius
g → profile exponent

preform
Glass rod, composed of → core and → cladding glass, which can be drawn out to form a → fiber.

probability of failure
Probability that an optical fiber of the length L will fracture due to a stress σ in the time t. The → Weibull distribution can be used to describe it.

profile exponent
Parameter which defines the shape of the profile for → power law index profiles. For practical use the profiles with an exponent $g \approx 2$ (→ graded index profile) and $g \to \infty$ (→ step index profile) are especially important.

profile dispersion
→ Refractive index of → fused silica glass depends on the → wavelength of the → light, but not in exactly the same way for all types of glasses used in an → optical fiber. The shape of the → refractive index profile (in particular the → refractive index difference) is therefore also dependent on the wavelength. For this reason the profile of → multimode fibers can only approach the optimum in a narrow wavelength range, which permits minimal → modal dispersion and maximum → bandwidth. At other wavelengths, the refractive index profile is not optimum and the bandwidth is correspondingly smaller.

pulse code modulation (PCM)
Modulation by which telecommunication signals are transmitted by means of → coding in the form of pulses. Analog telecommunication signals are digitized, coded and then transmitted digitally as a series of pulses.

Q

quantum efficiency
In a → transmitter diode, the ratio of the number of emitted photons to the number of charge carriers transmitted through the pn junction in the → semiconductor.

R

radial mode number
Number for characterizing the propagative → modes in an → optical fiber. The radial mode number gives the number of concentric → light rings of the mode (Figure 2.9).

radius of curvature
The curvature of a spatial curve is described by the radius of curvature or bending radius. For a → stranding element in a cable with a → lay length S and a → stranding radius R the radius of curvature can be calculated:

$$\varrho = R \cdot \left[1 + \left(\frac{S}{2 \cdot \pi \cdot R} \right)^2 \right]$$

Raman scattering
→ Scattering of → light in a medium according to the Raman effect, a nonlinear optical effect, whereby intensive → light radiation creates a broad continuous spectrum.

Rayleigh scattering law
→ Scattering due to density perturbations (inhomogeneities) in the → optical fiber, the dimensions of which are mostly smaller than the → wavelength of the → light, can be explained with close approximation by this law. As the wavelength increases, the scattering losses decrease with the fourth power.

receiver
Component in → optical communication technology for converting optical signals into electrical signals. It is composed of a receiver diode (→ PIN photodiode or → avalanche photodiode) with a → pigtail, a → connector and also a low noise amplifier and electronic switches for signal processing. The main parts of the receiver are usually combined to form a compact subgroup, the receiver module.

receiver sensitivity
Optical power needed by the → receiver for low error signal transmission. In digital signal transmission, usually the average optical power is stated in W or dB · *m* at which a → bit error rate of 10^{-9} is attained.

reflection
Effect in which rays (waves) are radiated back from the interface between two different media. In this effect the incident light ray and the reflected light ray lie in the same plane and their angles of incidence and reflection are identical. When → light passes from a medium with low → refractive index into a medium with a higher refractive index, only a portion of the light is reflected, the remainder is refracted or absorbed.

refracted near field method
Method for measurement of the → refractive index profile of an → optical fiber.

refraction
Change in direction of a ray (wave) when it passes from one medium into another and the → refractive indices of the two media are different, or in a medium, where the refractive index changes continuously as a function of the location (→ graded index profile). The angle of the incident and refracted rays can be calculated according to → Snell's law.

refractive index
Factor by which the speed of → light in an optically dense medium (e.g. glass) is smaller than it is in a vacuum. → Phase refractive index and → group

refractive index are differentiated. The refractive index in an optical medium, except in a vacuum, is a function of the → wavelength. It is also called the index of refraction.

refractive index contrast
→ refractive index difference.

refractive index difference
Difference between the highest refractive index occurring in the → core of an → optical fiber n_1 and in the → cladding n_2. The refractive index difference is the determining factor in the size of the → numerical aperture *NA* and the amount of additional → attenuation due to → microbending. The normalized refractive index difference is defined by

$$\Delta = \frac{n_1^2 - n_2^2}{2 \cdot n_1^2} \approx \frac{n_1 - n_2}{n_1}.$$

refractive index profile
Curve of the → refractive index n over the cross-section of an → optical fiber ($n = n(r)$).

repeater spacing
Optical cable distance between two regenerative repeaters.

reverse lay stranding
Method of → stranding, in which the → stranding elements of a stranding layer reverse direction after a certain number of turns along the length of the axis of the cable thereby creating an alternating S- or Z-shaped pattern.

ribbon cable
Cable in which → optical fibers are ordered in ribbons. Thereby the optical fibers are aligned parallel to each other with even spacing and glued in position, for example, between two adhesive foils. Several ribbons can be arranged as a stack on top of each other.

ring network
Network in which the individual stations (subscribers) are connected in a ring. They are not centrally controlled; the signals pass through each of the stations in series.

rod-in-tube method
One of the original methods for production of a → preform for → optical fibers. An ultrapure → fused silica glass rod as the → core is slid into a fused silica glass tube with lower → refractive index than the → cladding, the dimensions of the rod being such that hardly any free space remains. This method is only suitable for optical fibers with → step index profile.

S

scattering
Main cause of → attenuation in an → optical fiber. It is caused by microscopic density perturbations in the glass, which defect a part of the guided → light so far off its direction that it leaves the optical fiber. The scattering increases toward shorter → wavelengths according to the → Rayleigh scattering law (→ Raman scattering).

scattering constant
Proportionality constant between the → attenuation coefficient due to → scattering and the fourth power of the → wavelength of the scattered → light (→ Rayleigh scattering law).

semiconductor
Material used for electronic components. Semiconductors, e.g. germanium and silicon, are the basis both for individual transistors and for integrated circuits.

shear modulus
Proportionality constant between the shear of a body without change in volume and the force which acts in the direction of the side surfaces (tangential force per unit surface) (→ Poisson's ratio).

silicon dioxide (SiO_2)
Chemical compound of quartz or → fused silica glass. SiO_2 is the basic raw material in today's → optical fibers.

silicon tetrachloride ($SiCl_4$)
Highly volatile compound. In combination with oxygen it is used to produce ultrapure → silicon dioxide, the base material for today's → optical fibers.

single-fiber loose buffer
Composed of one → optical fiber and a surrounding loose buffer.

single-mode fiber
→ Optical waveguide which allows the propagation of only one mode, the → fundamental mode, at its operating wavelength.

slotted core cable
Cable in which the → optical fibers are not inside buffers but rather in preformed helical grooves in the surface of the central member.

Snell's law
When light is refracted both the incident → light ray and the refracted light ray lie in one plane, which is perpendicular to the interface of the two media. The ratio of the → refractive indices of the two media is equal to the ratio of the sine of the angle of incidence α to that of the refractive angle β:

$$\frac{\sin\alpha}{\sin\beta} = \frac{n_2}{n_1}.$$

sodium borosilicate glass
Multicomponent glass which is used as a base material for production of → optical fibers. Due to impurities, it exhibits high → attenuation.

softening point
Temperature at which a body changes its form due to its own weight. For → fused silica glass the softening point is 1730 °C, whereby the → viscosity $\eta = 10^{7.6}$ dPa · s or $\log \eta = 7.6$.

splice
Permanent connection of two mirror fractured → optical fibers, which can be produced with adhesives or by fusion.

splitter
Fiber optic component used to divide the light power from an incoming → optical fiber to two or more outgoing optical fibers (→ coupler).

spontaneous emission
Occurs when too many electrons are in the conduction band of a → semiconductor. These drop spontaneously into free spaces in the valence band, whereby one photon is emitted for each electron. The resulting radiation is incoherent.

star coupler
Fiber optic component which provides an even distribution of optical power between an identical number of incoming and outgoing → optical fibers. There are both passive and active star couplers.

star network
Network in which each station (subscriber) has its own line to the active branch point (→ star coupler).

steady state
Due to → mode mixing in a → multimode fiber, equilibrium mode distribution (EMD) of the → light power is reached after a certain distance, which does not change any more at longer distances. To approximate this steady state for measuring purposes, a → mode mixer and/or → mode filter are used.

step index profile
→ Refractive index profile of an optical fiber which is characterized by a constant refractive index within the core and a sharp decrease in the refractive index at the interface between → core and → cladding (→ profile exponent $g \rightarrow \infty$).

stimulated emission
Occurs in a → semiconductor when photons excite excess charge carriers in it to radiative recombination, i.e. excite them to emit photons. The → light emitted is identical in → wavelength and phase to the incident light, i.e. coherent (the radiation generated is coherent).

strain point
Lower limit of the → transformation range of a body. In → fused silica glass the strain point is 1075 °C, whereby the → viscosity $\eta = 10^{14.5}$ dPA · s or $\log \eta = 14.5$.

stranding
Twisting of → stranding elements. A differentiation is made between → helical stranding and → reverse lay stranding.

stranding element
In a cable these are basically → single-fiber loose buffers, → multifiber loose buffers, → tight buffered fibers and → composite buffered fibers, as well as fillers, copper pairs and copper quads.

stranding radius
Radial distance between the axis of the cable and the middle of a → stranding element.

system bandwidth
→ Bandwidth of an → optical fiber route segment measured from → transmitter to → receiver (→ repeater spacing).

SZ stranding
→ reverse lay stranding.

T

target rod
Rod of → fused silica glass or graphite onto which layers of oxide particles (soot) are deposited in the OVD (→ outside vapor deposition) method.

thixotropic
Property of certain gels that liquify when stirred and solidify when they are at rest again.

threshold current
Electrical current, above which the amplification of a → light wave in a → laser diode is greater than the optical losses, so that → stimulated emission begins. The threshold current is strongly temperature dependent.

through power technique
Method of measuring → attenuation in an → optical fiber. Two variations are differentiated: the → cut-back method and the → insertion loss technique.

tight buffered fiber
Buffered → optical fiber in which the firm plastic buffer is applied directly over the → coating.

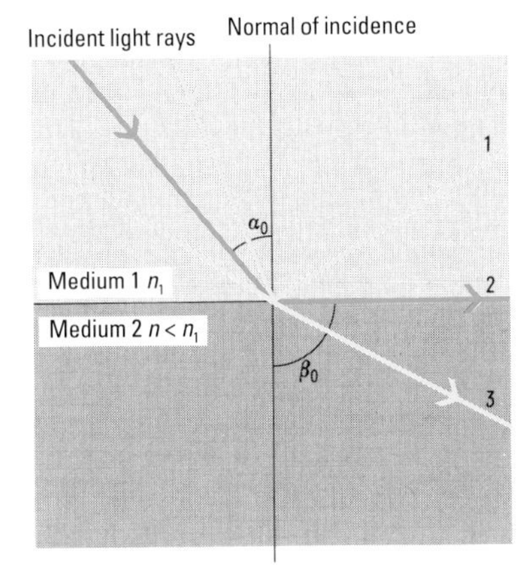

1 Totally reflected light ray
2 Refracted light ray
with angle of refraction $\beta_0 = 90°$
3 Refracted light ray

time division multiplex
Transmission method in which several parallel incoming digital signals are united into an outgoing temporally interleaved serial data flow.

total internal reflection
→ Reflection of all → light without → refraction when a light ray in an optically dense medium is incident to the interface with an optically less dense medium, whereby the angle of incidence must be greater than the → critical angle.

transfer function
An → optical fiber has the effect of a low-pass filter on the signals to be transmitted; i.e. it permits low frequency signals to pass and attenuates those with increasingly greater frequencies. This process is described by the transfer function $H(f_m)$. For this purpose the amplitude of the light power at the input $P_1(f_m)$ and at the output $P_2(f_m)$ of the optical fiber are measured for each → frequency of modulation f_m and the ratio of these amplitudes is calculated; the magnitude of the transfer function $H(f_m)$ is then

$$H(f_m) = \frac{P_2(f_m)}{P_1(f_m)}.$$

Its maximum value is at the zero frequency and it tends slowly towards zero for high frequencies. It approximates a → Gaussian low-pass filter.

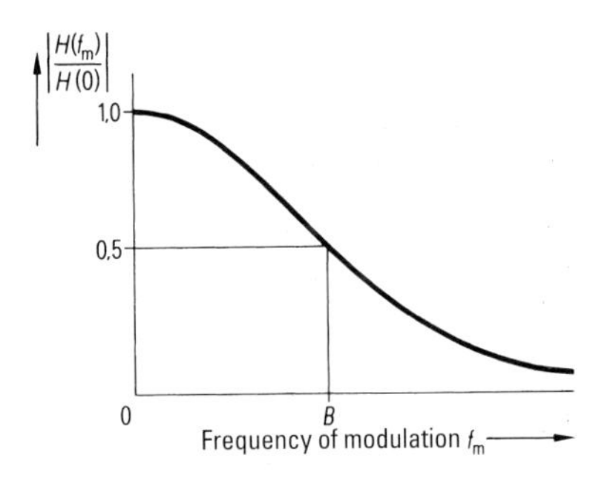

transformation range
Temperature range at which a body is transformed from the → viscoelastic state of a super cooled melt to the brittle state. In → fused silica glass this is between 1075 °C and 1180 °C.

transmitter
Component in → optical communication technology for converting electrical signals into optical signals. It consists of a transmitter diode (→ laser diode or → light emitting diode) with → pigtail, → connector and amplifier and other electronic circuits. In particular, in the case of laser diodes a → photodiode with regulating amplifier is needed to monitor and stabilize the radiative power; furthermore, in many cases a temperature probe and a Peltier cooler are needed for stabilization of the operating temperature. The main parts of a transmitter are assembled to a compact subunit, the transmitter module.

transverse offset method
Method for measurement of the → mode field diameter in → single-mode fibers.

transverse wave
Wave in which the oscillation takes place perpendicular to the direction of propagation (e.g. → electromagnetic wave in a vacuum). In contrast to it, the amplitude of longitudinal waves oscillates in the direction of propagation (i.e. sound waves).

U

ultraviolet radiation
Range of the spectrum of → electromagnetic waves from about 380 nm (violet) to about 10 nm. This coincides with the range of X-rays starting with 100 nm.

unit cable
Cable in which the → cable core is made up of units of stranded elements.

vapor axial deposition (VAD) method
Method for manufacture of → optical fibers by glass deposition out of the vapor phase onto the end face of a → fused silica glass rod.

viscoelastic condition
State of a body in which a deformation caused by an external force is only slowly reversed, when the force ceases to act on it, due to its → viscosity.

viscosity
Tendency to resist deformation due to internal friction; it is given in units of decipascal-seconds, whereby

$$1 \mathrm{dPa} \cdot \mathrm{s} = 1 \frac{\mathrm{g}}{\mathrm{cm} \cdot \mathrm{s}}.$$

In → fused silica glass the viscosity decreases continuously with increasing temperature T (Figure 3.1).

V number
Dimensionless parameter, dependent on the → core radius a, the → numerical aperture NA and the → wavelength λ or → wave number k of the → light. For example, it serves as a parameter for calculation of the number of → modes guided in the core of an → optical fiber:

$$V = 2\pi \frac{\alpha}{\lambda} NA.$$

Waveguide dispersion
→ Dispersion which describes the dependence of the group delay times of an individual → mode on the dimensions of the → optical waveguide and the → wavelength (in the case of material parameters not dependent on the wavelength). It is particularly important in → single-mode fibers, in which it is caused by the wavelength-dependent distribution of the → light of the → fundamental mode over → core and → cladding glass.

wavelength
Spatial period of a plane wave, i.e. the distance or path length of one complete oscillation. In → optical commu-

nication, wavelengths in the range of 0.8 to 1.6 μm are mainly used.

wavelength division multiplex (WDM)
Transmission method by which several units of information are transmitted simultaneously at different wavelengths over one → optical fiber and are separated by wavelength-dependent filters.

wave number
Phase shift of a wave per unit of length. The wave number k is inversely proportional to the wavelength λ with the factor 2π:

$$k = \frac{2\pi}{\lambda}.$$

Weibull distribution
Mathematical statistical method by which the → probability of failure can be determined, e.g. of an → optical fiber as a function of its length, of the mechanical stress and of the time.

Y

Young's modulus
Proportionality constant between the strain (elongation or contraction) of a body and the stress causing it (tensile force or pressure). This constant is also known as the modulus of elasticity (→ Hooke's law).

Bibliography

Books

Adams, M.J.: An Introduction to Optical Waveguides. Chichester, New York: John Wiley & Sons 1981

Agrawal, G.P.: Fiber-Optic Communication Systems. New York: John Wiley & Sons, 2nd edition 1997

Agrawal, G.P.: Nonlinear Fiber Optics, Academic Press: 3rd edition 2001

Akhmediev, N.N.; Ankiewicz, A.: Solitons. London: Chapman & Hall 1997

Arnaud, J.A.: Beam and Fiber Optics. New York, London: Academic Press 1976

Barnoski, M.K.: Fundamentals of Optical Fiber Communications. New York, London: Academic Press, 2nd edition 1981

Becker, P et al.: Erbium-Doped Fiber Amplifiers. Academic Press 1999

Best, S.W.: Nachrichtenübertragung mit Lichtwellenleitern. Heidelberg: Hüthig 1983

Börner, M.; Trommer, G.: Lichtwellenleiter. Stuttgart: Teubner 1989

Crisp, J.: Introduction to Fiber Optics. Butterworth-Heinemann 1997

Deutsch, B.; Mohr, S.; Roller, A.; Rost, H.: Elektrische Nachrichtenkabel. Erlangen: Publicis MCD 1998

Elion, G.; Elion, H.: Fiber Optics in Communications Systems. New York: Marcel Dekker 1978

Faßhauer, P.: Optische Nachrichtensysteme. Heidelberg: Hüthig 1984

Geckeler, S.: Lichtwellenleiter für die optische Nachrichtenübertragung. Berlin: Springer, 3. Aufl. 1990

Glaser, W.: Lichtwellenleiter, Eine Einführung. Berlin: Verlag Technik, 3rd edition 1990

Grau, G.: Optische Nachrichtentechnik. Berlin: Springer 1986

Harth, W.; Grothe, H.: Sende- und Empfangsdioden für die optische Nachrichtentechnik. Stuttgart: Teubner 1984

Hecht, J.: Understanding Fiber Optics. Prentice Hall, 3rd edition 1999

Herter, E.; Graf, M.: Optische Nachrichtentechnik. München: Hanser 1992

Heinlein, W.: Grundlagen der faseroptischen Übertragungstechnik. Stuttgart: Teubner 1985

Jeunhomme, L.B.: Single-Mode Fiber Optics. Principles and Applications. New York, Basel: Marcel Dekker 1983

Kao, C.K.: Optical Fiber Systems: Technology, Design, Applications. New York: McGraw Hill 1982

Kaminow, I.P.: Koch, T.L.: Optical Fiber Telecommunications IIIA, IIIB: Academic Press 1997

Kersten, R.Th.: Einführung in die optische Nachrichtentechnik. Berlin: Springer 1983

Li, T.: Optical Fiber Communications, Band 1: Fiber Fabrication. Orlando: Academic Press 1985

Marcuse, D.: Light Transmission Optics. New York: Van Nostrand Reinhold, 2nd edition 1982
Marcuse, D.: Theory of Dielectric Optical Waveguides. New York, London: Academic Press 1974
Marcuse, D.: Principles of Optical Fiber Measurements. New York, London: Academic Press 1981
Midwinter, J.E.: Optical Fibers for Transmission. New York, Chichester: John Wiley & Sons 1979
Miller, C.M.: Mettler S.C.; White I.A.: Optical Fiber Splices and Connectors. New York, Basel: Marcel Dekker 1987
Murata, H.: Handbook of Optical Fibers and Cables. New York, Basel: Marcel Dekker 1988
Okamoto, K.: Fundamentals of Optical Waveguides. Academic Press 2000
Okoshi, T.: Optical Fibers. New York, London: Academic Press 1982
Palais, J.C.: Fiber Optic Communications. Prentice Hall, 4th edition 1997
Personick, S.D.: Optical Fiber Transmission Systems. New York, London: Plenum Press 1981
Personick, S.D.: Fiber Optics, Technology and Applications. New York, London: Plenum Press 1985
Rosenberger, D.: Optische Informationsübertragung mit Lichtwellenleitern. Grafenau/Berlin: Expert-Verlag/VDE-Verlag 1982
Sandbank, C.P.: Optical Fiber Communications Systems. Chichester, New York: John Wiley & Sons 1980
Schubert, W.: Nachrichtenkabel und Übertragungssysteme. Berlin, München: Siemens AG, 3. Aufl. 1986
Senior, J.M.: Optical Fiber Communications. London: Prentice-Hall 1985
Sharma, A.B.: Optical Fiber Systems and their Components. Berlin: Springer 1981
Shotwell, R.A.: An Introduction to Fiber Optics. Prentice Hall 1997
Snyder, A.W.; Love, J.D.: Optical Waveguide Theory. London: Chapman & Hall 1983
Sodha, M.S.; Ghatak, A.K.: Inhomogeneous Optical Waveguides. New York: Plenum Press 1977
Suematsu, Y.; Iga, K.I.: Introduction to Optical Fiber Communications. New York, Chichester: John Wiley & Sons 1982
Unger, H.G.: Optische Nachrichtentechnik. Berlin: Elitera 1976
Unger, H.G.: Optische Nachrichtentechnik. Bd. 1: Optische Wellenleiter, Bd. 2: Komponenten, Systeme. Meßtechnik. Heidelberg: Hüthig 1984
Unger, H.G.: Planar Optical Waveguides and Fibers, Oxford: Clarendon Press, 2nd edition 1980
Weinert, A.: Plastic Optical Fibers. Erlangen: Publicis MCD 1999
Winstel, G.; Weyrich, D.: Optoelektronik I, Lumineszenz- und Laserdioden. Berlin: Springer 1980
Yeh, C.: Handbook of Fiber Optics. San Diego, London: Academic Press 1990

Conference Proceedings

European Conference on Optical Communications (ECOC). London 1975, Paris 1976, Munich 1977, Genova 1978, Amsterdam 1979, York 1980, Copenhagen 1981, Cannes 1982, Geneva 1983, Stuttgart 1984, Venice 1985, Barcelona 1986,

Helsinki 1987, Brighton 1988, Gothenburg 1989, Amsterdam 1990, Paris 1991, Berlin 1992, Montreux 1993, Florence 1994, Brussels 1995, Oslo 1996, Edinburgh 1997, Madrid 1998, Nice 1999, Munich 2000, Amsterdam 2001

International Conference on Integrated Optics and Optical Communications (IOOC). Tokio 1977, Amsterdam 1979, San Francisco 1981, Tokio 1983, Venice 1985, Reno 1987, Kyoto 1989, Paris 1991, San Jose 1993, Hongkong 1995, Edinburgh 1997, San Diego 1999

International Wire & Cable Symposium (IWCS). Asbury Park 1952 until 1963, Atlantic City 1963 until 1974, Cherry Hill 1975 until 1983, Reno 1984, Cherry Hill 1985, Reno 1986, Arlington 1987, Reno 1988, Atlanta 1989, Reno 1990, St. Louis 1991, Reno 1992, St. Louis 1993, Atlanta 1994, Philadelphia 1995, Reno 1996, Philadelphia 1997 and 1998, Atlantic City 1999 and 2000, Orlando 2001 and 2002

Optical Fiber Communication Conference (OFC). Williamsburg 1975 and 1977, Washington D.C. 1979, San Francisco 1981, Phoenix 1982, New Orleans 1983 and 1984, San Diego 1985, Atlanta 1986, Reno 1987, New Orleans 1988, Houston 1989, San Francisco 1990, San Diego 1991, San Jose 1992, 1993 and 1994, San Diego 1995, San Jose 1996, Dallas 1997, San Jose 1998, San Diego 1999, Baltimore 2000, Anaheim 2001

Index